PRIVATE PILOT FLIGHT TRAINING MANUAL

SECOND EDITION

Written by

Ralph Butcher

PUBLISHER AND DISTRIBUTER
Skyroamers® Publications
www.skyroamers.com
870 North Fern Street, Orange, CA 92867-7235

Printed and bound in the United States of America.

INTENT

To provide a practical training manual that belongs in the cockpit, not the classroom. A manual that concentrates on flying the airplane—standard operating procedures, flight maneuvers, and operational insights.

APPLICABILITY

This publication does not amend, supersede or take precedence over any regulation, directive, or procedure mandated by either the Federal Aviation Administration or the airplane's pilot's operating handbook.

COLOPHON

This manual was produced on an Apple computer using Adobe's FrameMaker™ and Illustrator™ software. Book designed and illustrated by the author. Cover illustration: Beechcraft A-36 with permission from Beech Aircraft Corporation. Printed by Crestec Los Angeles, Inc.

ISBN 1-881688-04-6 34.95

Library of Congress Control Number: 2004090822

TABLE OF CONTENTS

FOR SKYROAMERS® INFORMATION AND ORDERING
GO TO WWW.SKYROAMERS.COM

ILLUSTRATIONS

INTRODUCTION

This *Skyroamers*® publication is a practical manual that simplifies the art of flying. It was written for the cockpit, not for the classroom. With the exception of chapter 9, all references pertain to fixed-gear, fixed-pitch propeller airplanes.

I could have published this manual with a fancy multicolored cover and lots of inspirational airplane photos, but that's not my style. I want my words to be the focal point of your purchase, not the artsy whim of a non-flying book designer. I'll let my years of experience as a commercial pilot and an innovative flight instructor be the return on your investment. You'll enjoy the results. I guarantee it.

In 1959, the FAA introduced the concept of *integrated flight instruction* in order to add instrument flight training to basic, visual flight training procedures, an action that was intended to reverse the increased rate of instrument-related accidents by instrument-rated pilots. General aviation airplanes and avionics had become more sophisticated, and more pilots were acquiring instrument ratings.

As so often happens, solving one problem generates another, and integrated flight instruction did just that. It caused an increase in stall-spin accidents, because pilots were now trained in a manner that developed an overreliance on instrument reference. The art of using visual references for visual flying was de-emphasized.

Integrated instruction produced a generation of airplane drivers, not pilots. Airplane drivers cannot fly the visual maneuvers without instrument reference, they are unaware of the proper visual references, and their flying is very mechanical. They never mastered—and were probably never introduced to—the proven, basic visual flying techniques that were taught in this country for many years. That unwise, inexcusable flight training error has caused accidents.

With the introduction of ground training devices—personal computer aviation training devices, flight training devices, and flight simulators—a new solution to the original problem emerged. With these devices—particularly those that have the aerodynamic responses of an airplane—a knowledgeable instructor can easily meet and exceed the intent of integrated instruction when training pilots for the instrument rating.

Because of that technology, I reverted back to the proven visual training procedures that preceded integrated instruction. My training syllabuses and this manual support that objective by teaching you how to fly the wing—the common denominator for all airplanes—and how to use the proper visual references.

When I introduce a new maneuver to a student, the flight instruments are covered. After the student understands the maneuver and can fly it using visual references and the concept of *attitude plus power equals performance*, I introduce instrument reference in order to achieve the required level of preciseness. When flying in visual conditions, instrument reference is always the third priority, never the first.

My approach to flight training has been well received by the many instructors and students who use my publications. I thank them for their support.

This edition of the manual continues to follow my training philosophy of maximum confidence, minimum workload. And remember, that while it is impossible to teach experience, I can and have used my extensive flight experience to simplify what is taught.

The following article is one of many that I wrote as a contributing editor for the AOPA's *Flight Training* magazine. It explains my extremely successful flight training philosophy.

Common Denominators

Avoid the information dilemma - April 2003 issue

"I'm a student pilot, and I have a problem. If I ask two flight instructors and an FAA friend a question about flying, I get three different opinions. The same thing happens when I ask a corporate or airline pilot or when I read magazine articles. Will someone please tell me what's correct?"

At one time or another, we have all been in the above situation. To resolve the information dilemma, you must consider two common denominators: operating mode and training philosophy.

Operating mode refers to aircraft and crew. For initial, general aviation flight training, the operating mode is one pilot in a light, piston-powered airplane. The procedures taught at this level are the prerequisites for the advanced operating modes—piston-powered twins, turboprops, jets, and two-pilot crews. This evolutionary process, however, does not work in reverse. What pilots do in the advanced operating modes does not apply to initial, general aviation flight training.

The first step for reducing the information dilemma is to determine the operating mode for the pilot who is conveying the new information. If you both have the same mode, that information may have merit.

The second step is more difficult. You must understand the training philosophy that the pilot in question was trained under. There are many variations. The *pass the test* philosophy is the worst case—the pilot was prompted using actual written test questions and trained to pass an examiner's flight test, the details of which were well known to the instructor. The next worst is the *cram course*—the proper mental and physical habit patterns are seldom established. Six months after course completion, proficiency and knowledge vaporize.

At the top end of the spectrum is the *maximum confidence and minimum workload* training philosophy that I developed years ago. Students trained in this manner can usually separate fact from fiction with confronted with the information dilemma.

Building maximum student confidence is a critical flight instructor responsibility. I do this by minimizing the number of procedures that are taught, but to qualify, each procedure must work in the worst possible situation that a pilot will encounter when flying. Two additional advantages occur: learning becomes easier and repetition increases. Repetition is the key element for developing ingrained habit patterns.

One example of a worst-situation procedure is the constant attitude and power, variable airspeed approach, developed years ago by Wolfgang Langewiesche, author of *Stick and Rudder.* Students who are taught this technique master forced-landing-thinking during presolo instruction, and later on, they have no difficulty with night landings, particularly when surface lighting causes optical illusions.

A fine line separates self-confidence and overconfidence. Self-confidence occurs when students follow procedures that will keep them out of trouble, know their limitations, and use good judgment. Overconfidence occurs when students make erroneous assumptions regarding their attitude, knowledge, or proficiency.

Minimum workload means that students are taught to maximize their free time in the cockpit so that emergencies, unusual situations, and distractions can be handled without compromising basic flying skills or flight safety. Passengers, watching a properly trained pilot at work, usually think that flying is easy. Unfortunately, they can't see the wheels that are spinning inside the pilot's head.

If two pilots perform the same task in 30 and 50 seconds respectively—this does not mean that one pilot is flipping switches as fast as he can—the second pilot should change his method. However, if both pilots perform the task in 30 seconds, each using a different method, all is well. Individual variations are perfectly normal in this case.

You can avoid the information dilemma by applying common denominators in whatever you read, see, or hear. What is the operating mode? Will the procedure or technique work in the worst situation that you will encounter? Will the procedure or technique reduce the workload?

— — — — —

Thank you for purchasing my manual. It's the entry key to the wonderful and extremely enjoyable world of flying. If you ever have a question regarding the content, please e-mail your question to ralph@skyroamers.com.

Sincerely,

Ralph Butcher

AOPA's *Flight Training* magazine's motto:

A GOOD PILOT IS ALWAYS LEARNING!

CHAPTER 1

PILOT TRAINING

HOW TO START

There are many excellent flight instructors and flight schools in our country.

Good luck! There are many excellent flight instructors and flight schools in our country, but some should not be in business. Do not start flight training until you do some research. This chapter discusses several pertinent factors.

To start, you should ask other pilots that you know for recommendations. Next, contact your local Federal Aviation Administration (FAA), Flight Standards District Office (FSDO—say *fiz-doe*). The FSDO has a list of pilot examiners who conduct pilot flight testing and certification for the FAA. These examiners—particularly the independent ones who are self-employed— know the best schools and the best instructors.

To receive a referral that takes you into consideration, arrange to meet the examiner in person if at all possible. Then schedule a short meeting with each recommended instructor, and ask a few general questions about their background and why they enjoy teaching. Look for a person who communicates well, is sincere, shows enthusiasm, and has an appearance and demeanor that puts you at ease. Finally, the instructor and school should have the attributes that are listed in the following sections.

FLIGHT INSTRUCTORS

Flight instructors have all the variables of the human race—male, female, young, old, tall, short, skinny, fat. Some are building experience and working toward advanced ratings, others just enjoy teaching and helping pilots progress up the certification ladder. All of them love to fly and share the camaraderie that flying generates.

Anyone can instruct, but can they teach? That is the key question. Teachers are identified by the following traits:

- Communicate well—they have a gift for making the complex seem simple, and they know how to teach the basics.

- Never shows impatience.

Teachers recognize errors and give solutions.

- Identify and correct student errors. Any instructor can point out errors, but teachers give solutions.

- Determine if students have the judgment and character required of a competent pilot. Some individuals are not suited for this endeavor and should not be trained.

- Never hedge a question or pretend to know everything. If unsure they will admit it and find the correct answer.

- Are enthusiastic, encouraging, positive, and show sincere student interest.

- Use a flight training syllabus and accurately record all training that students receive.

- Are on time and well prepared for each lesson.

- Are appropriately dressed and well groomed.

Teachers charge for all ground and flight instruction.

- Charge students for all ground and flight instruction. This includes the preflight and postflight briefings.

- Never cancel students in order to fly a charter trip.

- Cancel the airplane and give ground instruction when students are not prepared for a flight lesson.

• Drop students who repeatedly forget or are late for scheduled lessons or who do not apply themselves.

Teachers let students do most of the talking.

• Talk when explanation must be given, but primarily ask questions and let the student do most of the talking. Good instructors know that students retain 10% of what is read, 20% of what is heard, 30% of what is seen, 50% of what is seen and heard, 70% of what they say, and 90% of what they say while performing that task. [William Kershner, *The Flight Instructor's Manual*]

• Take extra time at the beginning of the course to determine the student's learning characteristics and how the student responds to various instructor inputs—loud voice, soft voice, criticism, etc. With these early insights, a teacher can optimize the critical teacher-student relationship.

• Carry a note pad in the airplane to reference student performance. Major items are mentioned in flight; minor items are held for the postflight briefing.

• Carry half of a manila file folder in order to cover up the flight instruments when required. The reason and method for doing this is explained in chapter 3.

Teachers never allow students to use instrument reference or radio navigation until the basics are mastered.

• Never allow students to use instrument reference or radio navigation until the basics are mastered. You master the art of flying by using outside attitude references and the concept of attitude plus power equals performance; you master navigation by using land-fall, pilotage, and dead-reckoning procedures. Then and only then are advanced concepts introduced.

• Teach the mechanics of each maneuver before they teach preciseness. They know that proficiency is not synonymous with preciseness. Proficiency evolves from a full command of basic fundamentals. Preciseness evolves from practice—the repetitive application of basic fundamentals.

• Never teach visual reference points that apply only to a specific airplane or airport. Teachers use references that apply to all airplanes and all airports.

• Let students make mistakes and thereby learn from actual experiences. When you see or hear a student make an error, do not laugh. There may be an outstanding teacher sitting in the instructor's seat.

• Teach students to cope with distractions during each phase of flight.

To demonstrate final proficiency, each maneuver must be correctly performed on three separate occasions.

• Insist that students correctly perform each maneuver and skill on three separate occasions before they advance too far in the syllabus. Correct performance on one occasion is not enough. Students must continue to review the task until correct execution is habitual. A maneuver is mastered only when consistent performance has been demonstrated.

• Spend a portion of each lesson reviewing previous maneuvers because repetition maximizes retention.

• Fly the airplane only when absolutely essential for the purpose of demonstrating a maneuver.

• Assume physical control of the airplane only when real trouble is imminent.

• Never touch a switch or control on the instrument panel unless absolutely necessary. If something is incorrect, a teacher will make subtle suggestions to the student.

Teachers always follow standard operating procedures.

• Follow standard operating procedures (SOPs). Instructor credibility decreases and student apathy increases when a student pilot observes an instructor violate SOPs.

• Work vigorously at making students respect the responsibilities that are associated with piloting an airplane.

• Maintain student enthusiasm by ending each lesson on a positive note. Most lessons will end with an example that shows student progress.

• Have students record their lessons on cassette tapes whenever possible for review and study purposes.

• Utilize another instructor for the stage checks that are specified in the training syllabus.

• Inform students of their training potential within the first ten hours of training. Do they have the ability to become a pilot? Will training time be longer than normal?

• Conduct stage checks as a flight lesson and not just a test when checking another instructor's student. Yes, they are checking, but when student performance can be enhanced, a teacher does so.

Stage checks are an important part of the training process.

The importance of stage checks cannot be overlooked, because all students encounter occasional difficulties during training. They do not always ask the right questions, and instructors do always have perfect explanations. This can leave the student with some reservations about a maneuver or a knowledge element. To rectify this situation, a stage check instructor will pay particular attention to student problems that occurred during that stage.

Instructors convey considerable information to their students, and the importance of some elements tends to diminish over time. The stage check instructor performs the vital function of reinforcing the critically important elements.

When two independent instructors—they are not affiliated with a flight school—work with each other's students as check instructors, student proficiency always increases. It also increases when more than one student is taken on a training flight. The observer always benefits from watching the student in the pilot's seat.

• Insist that employers and aircraft owners have adequate insurance for the instructor.

• Accept a student's request for an instructor change.

• Change a student to another instructor when it is obvious that personalities are incompatible.

Teachers use my training philosophy of *maximum confidence and minimum workload.*

• Never takes the position of "Do it my way or else." Teachers use my *maximum confidence and minimum workload* training philosophy that is explained in the introduction to this manual. This philosophy lets instructors immediately resolve opinions that differ from recognized practices.

STUDENTS

Students must make several commitments.

A good flight instructor and unlimited money cannot guarantee success. To obtain a pilot's certificate or rating, students must make the following commitments:

- Make certain that they are medically qualified for their aviation goal before they start flight training. Aviation medical examiners conduct the required physicals. If you desire a professional flying career, I strongly recommend that you pass an extensive physical examination including an EKG from your personal physician and an extensive eye examination from an ophthalmologist.

- Study the required material before the lesson and review it after the lesson. Instructors cannot give you a pill that allows all learning to occur at the airport.

- Fly at least twice a week if they want to progress at a reasonable rate.

Ensure that money is available to cover training costs.

- Ensure that money is available to cover all training costs. Some schools and instructors make false statements regarding training time. Those students are then forced to curtail training which wastes previous expenditures. No one can give you an exact estimate. Currently, the national average for private pilot training is 70 hours of flight training plus the cost of ground training, supplies, and the FAA checkride.

- Set simple goals. Do not start off the first day dreaming about being a commercial pilot. Concentrate on soloing, then on soloing for cross-country flights, and finally on private pilot certification. Take training one step at a time, and obtain the most out of each stage of instruction.

Be on time for lessons and be well prepared.

- Are on time for lessons and are well prepared. Time and money are wasted when you are not mentally prepared for a lesson. If this occurs, the instructor will cancel the airplane and spend the scheduled time giving you ground instruction.

- Write down questions that come to mind at home in order to discuss them during the instructor's preflight briefing.

Do not fly when under external pressure.

• Do not fly when under external pressures of business, family, or illness. And if you hoot with the owls, you can't soar with the eagles. If a lesson must be cancelled, let your instructor know as soon as possible.

• Do not compare their progress with that of others. The learning process applies to individuals not groups. Students progress at their own pace.

• Must have a good command of the English language.

• Realize that progress does not occur with every lesson. There are times when the learning curve flattens out. Remember the old saying, "You take two steps forward and one step back."

• Do not intimidate the instructor or attempt to redirect the training program. This can occur when an older student has a young instructor; when that happens, the student always has problems in the long run.

• Dress appropriately and have acceptable personal hygiene. High-heeled shoes, garlic breath, and body odor have no place in small airplanes.

• Must not assume that one instructor is inferior to another. Age, sex, race, physical features, and experience does not reflect teaching ability.

• Must not assume that an experienced instructor is superior to a new instructor. I usually hire instructors with no experience providing they display the traits that I discussed in the previous section. They do exceptionally well.

Training time is directly related to the ability to study and retain information.

• Realize that length of training is directly related to the ability to study and retain information. Training time will increase for individuals who do not have good study habits, who have deep-rooted habit patterns that must be broken—age is a major factor—or who have a hectic schedule that puts flying at the bottom of their priority list.

• Must assess whether or not they truly enjoy flying after the first few hours of instruction. If not, do not continue.

• Must accept a change of instructors if recommended by the current instructor or request a new instructor if a personality conflict develops.

A good instructor-student relationship must be maintained.

• Must respect the instructor's abilities and desires in order to maintain a good instructor-student relationship. To be successful, students must make a genuine, sincere effort to comply with their instructor's requests.

GROUND TRAINING

For every hour you spend in an airplane, you must spend three to four hours studying.

In addition to actual flight training, considerable time must be spent studying and receiving ground instruction in order to master the subject—your first priority—and pass the FAA examinations—your second priority. For every hour you spend in the airplane, you must spend at least three to four hours studying.

Two examinations are required for pilot certification. The knowledge examination (written test), administered at FAA designated testing facilities, must be passed with a minimum grade of 70 percent. The practical examination (oral and flight test) are administered by an FAA designated pilot examiner (DPE).

Ground school training is essential.

Formal ground schools can be found at many flight schools and community colleges. If a formal ground school is not available, two options exist. The most effective is a home study course that provides textbooks, video or audio cassettes, and sample examinations. The least effective is studying only the appropriate textbooks. Your flight instructor will make specific recommendations.

If you decide to self-study, your flight instructor should give you a progress check after you complete each subject—aerodynamics, systems, weight and balance, performance, navigation charts, airspace, flight planning, meteorology, federal regulations, the Aeronautical Information Manual (AIM), and the Airport Facility Directory (AFD). Some areas are difficult to comprehend by just reading a book, but mastery of each subject is mandatory.

3-day ground schools are fast moving review sessions and should not be utilized as a primary source of training.

In some cities, 3-day ground schools are offered. These are fast moving review sessions and should not be utilized as a primary source of training. If you are unable to enroll in a formal ground school and are forced to use self-study, take the 3-day ground school after you feel you have properly prepared for the FAA written examination.

If you are using a Skyroamers® flight training syllabus, part one of that publication—the part that you keep—contains a lesson summary with study assignments and a supplemental study guide that makes it easier to learn the required federal regulations and AIM procedures.

FLIGHT TRAINING PROVIDERS

Where can you learn to fly? There are four types of providers: the military, universities and community colleges, flight schools, and independent instructors.

MILITARY

To have graduated from a military flight school opens every door in the aviation community. Nothing compares to this level of training, equipment sophistication, and mental conditioning. These graduates have proven themselves in every respect and are prime candidates for commercial flying jobs.

UNIVERSITIES AND COMMUNITY COLLEGES

For career insurance, obtain a college degree in a non-aviation field.

Several universities and community colleges offer flight training. Most programs are excellent, but because obtaining and maintaining a flying job is so unpredictable, I advise people to get a college degree in a field that they could pursue if a flying career does not work out due to business failures, layoffs, mergers, or personal medical problems.

FLIGHT SCHOOLS

Flight schools can be independent businesses or be associated with fixed-base operators (FBOs) and flying clubs. FBOs are commercial businesses that originally provided all general aviation services—aircraft sales, maintenance, charter, fuel, rental, and training—but in recent years, many have restructured and now concentrate on two or three fields of endeavor. Flying clubs charge various forms of membership fees, rent airplanes, and provide flight instruction.

Today, universities and community colleges do a large portion of our country's civilian flight training. Consequently, the FBO flight schools and flying clubs have decreased in number and size, but they still serve those who are not interested in the university or community college programs.

Most schools employ new flight instructors willing to work for low pay in order to accumulate the flight time required for higher paying jobs. Working at a school gives new instructors added prestige, provides a supply of students, qualifies them for reduced rates on airplanes that can be used for advanced ratings, and may provide an opportunity for charter flying.

Word of mouth advertising is a school's lifeline.

A successful school will charge a premium fee for instructors, airplanes, and services, and it will have managers with extensive flight training and piloting experience. I have never interviewed an insincere flight instructor, but I have interviewed many who were misinformed. Unfortunately, they were trained in an environment that did not provide proper leadership. A properly managed school standardizes their instructors and produces satisfied customers who readily support that business. Word of mouth advertising from these customers is a school's lifeline.

Most school and flying club aircraft are leased from an independent owner and are referred to as a leaseback. In a leaseback, the airplane owner pays for all operating expenses, and the school pays the owner a percentage of the rental fee. The owner may realize tax advantages if the airplane qualifies as a business investment.

The unhappiest leaseback owners are those who assumed that they could fly for free if their airplane was leased.

Leasebacks work well for some people but poorly for those who assumed that they could fly for free if their airplane was leased. This is seldom true even though the numbers can make it look attractive. If the owner cannot afford to pay for the airplane when the school's utilization is low, problems result, and many first-time owners do not fully understand aircraft maintenance requirements and expenses.

The insurance carrier is a school's nemesis. If the school's safety record is poor, increased insurance premiums will eventually shut down the business. To avoid a school with an uncertain future, renter pilots and students should not look for the cheapest price. The school with stable ownership and competent management is usually a good business to deal with, and yes, these schools place high demands on renter pilots and students.

Renter pilots must comply with specific FAA and flight school requirements.

Renting an airplane is not like renting a car. You must comply with FAA regulations and the flight school's dispatch requirements. A renter pilot who gets upset because a school's dispatcher will not allow him to depart on a planned dream vacation due to a lapsed certificate or currency requirement should probably not be flying in the first place.

The relationship between a school and a renter pilot requires mutual consideration. The school provides airplanes to pilots who, in most cases, have no other access to airplanes. In return, the pilot must comply with FAA and flight school operating rules, must show good judgment, and must understand the uncertainties of the aircraft rental business. A successful school will not rent airplanes to those who violate the rules or refuse proper training.

A major uncertainty is schedule dependability. A school can never guarantee that a specific airplane will be available as scheduled. Maintenance problems and unforecast weather are two examples that are impossible to predict. Good schools have conscientious dispatchers who attempt to resolve such problems and keep the next renter pilot informed of the situation so that alternate plans can be made.

It is impossible to guarantee the amount of flight time required to obtain a specific certificate or rating.

Responsible schools have a problem selling flight training when the competition unethically understates actual training time. Because of previously mentioned factors—quality, frequency, and location of training; and your background, age, motivation, and ability to study—it is impossible to guarantee the amount of flight time required to obtain a specific certificate or rating. In spite of this, many flight schools attempt to attract students by advertising the minimum training times listed in the regulations even though those training times are totally unrealistic in today's aviation environment.

Successful schools advertise actual student performance.

Successful schools advertise actual student performance. Student completion time is categorized for each certificate and rating by keeping separate statistics based on age, frequency of instruction, education and occupation. These records also help the school monitor instructor performance.

Good schools provide these services:

- A courteous, knowledgeable, and well trained staff capable of handling customers, leaseback owners, and maintenance operations in a responsible and ethical manner.

- Experienced instructors to meet the needs of experienced pilots undergoing advanced training, to work with problem students assigned to less experienced instructors, to conduct stage checks, and to ensure that new instructors are properly trained and standardized.

- Personnel who make certain that the airplanes are clean inside and out, full of fuel and oil, and contain the proper FAA documentation, checklists, and pilot's operating handbook.

- Dispatchers who anticipate customers' needs to include timely notification of medical certificate, flight review, and renter-currency expiration dates.

Flying once every three months does not ensure pilot proficiency.

- Require customers to maintain proficiency above FAA minimums. Flying once every 90 days does not ensure pilot proficiency. In the U.S. Army, which operated light airplanes, pilots with desk jobs would not get flight play unless they flew at least four hours per month. That's an excellent rule.

- Full disclosure of renter pilot insurance liability.

- Maintenance personnel that give full support to the flight school and leaseback owners.

- Newsletters and seminars that maintain customer interest in flight safety, proficiency training, and the school's operation.

- Comprehensive syllabi for each course of instruction so that all training can be documented and certified. That record remains on file with the school for liability protection.

- Comprehensive aircraft checkouts and flight reviews with appropriate documentation for liability protection.

Successful schools have structured programs, sound policies, competent employees, and good management.

If a school does not perform in the manner I have described, something is wrong. Be wary!

FAA flight school approval is not a guaranteed sign of superiority in relation to other schools. It signifies that certain FAA standards have been met and that training can be conducted under part 141 of the federal regulations. Independent flight instructors and other flight schools train under part 61. Pilots trained under either regulation must be equally proficient.

Parts 61 and 141 describe the requirements and minimum training time for each pilot certificate and rating. These minimum times, established years ago, have become obsolete due to today's advanced airplanes and complex operating environment. With minor exceptions for the commercial pilot certificate and the instrument rating, FAA approved flight schools cannot honestly say that their training programs will take less time than those offered by other flight schools even though this was the original intent of part 141.

The difference between FAA minimum and actual flight training times is due to a United Nations requirement. As a member of the UN, the United States must comply with certain International Civil Aviation Organization (ICAO) regulations, and FAA training times reflect the worldwide standard.

The FAA embraces the concept of demonstrated competency rather than flight time acquisition.

The FAA has addressed this problem by shifting the emphasis from training times to comprehensive testing standards. They are embracing the concept of demonstrated competency rather than flight time acquisition, and I applaud them for this improvement. Examiners can no longer be selective during flight testing. Each applicant must be tested in each operation that pertains to the certificate or rating being sought. If applicants meet specific practical test standards and show good judgment, they will pass. If not, additional training will be required.

INDEPENDENT INSTRUCTORS

Recommended, independent instructors are well worth any inconveniences caused by part time schedules or a waiting list. Those with considerable experience and recognized teaching abilities will charge premium rates for their services.

INSIGHTS (More of my articles from the AOPA's *Flight Training* magazine)

Sad But True

Not all training is equal - March 2004 issue

A recent, nationally-distributed article, *Why Students Switch Schools*, set off all my bells and whistles. The author stated this: "Many students switch flight schools in the middle of their training because of two vexing issues. One of the most common triggers is an unscheduled change of CFIs. Another common trigger is the requirement to fly with your chief pilot for an interim 'stage check' (even though they're just a Part 61 student) or for a precheckride flight check. These two fairly common events somehow make students feel like they are being ripped off and over charged. They switch schools."

Those statements are true. What irritates me is that such schools and CFIs allow this to happen, and they do not know how to handle these events properly.

Changing instructors Yes, CFI changes do occur, but they should never impede student progress. The first requirement for any flight school is to standardize their instructors and use a comprehensive flight training syllabus for all students.

Too many schools let their instructors do their own thing. The instructor is never critiqued unless a student causes an incident—hopefully not an accident—or doesn't pass a practical test on the third try. Yes, I said third try, not the second.

These schools use brief training outlines—many of which are produced commercially—that give little if any instructor guidance. Student documentation is limited to the instructor's cryptic notes in the student's log book. No record of student performance or the review process is kept, which makes it impossible for a new instructor to maintain training continuity.

When proper documentation exists, instructor changes—or substitutions due to vacations or illness—can occur without affecting student progress. The only caution area is personalities. The current flight instructor has the best insight regarding the student's personality and the personalities of the other instructors. That instructor should arrange the transfer; otherwise, it becomes the chief instructor's responsibility.

Stage checks These checks are essential for high-quality training, regardless of the regulation that the student is trained under—part 61 or 141. If the intent is to train a student to proficiency—the proper habit patterns have been fully developed—the training in both cases will be identical. If the intent is to train the student in minimum time for the FAA's practical test, the proper habit patterns have not been not fully developed, the student has been robbed, and flight safety has been compromised.

The FAA may disagree with me, but a properly executed stage check is not just a cold-hearted evaluation with a pass or fail result. A stage check should enhance the training program (I call it team teaching) and benefit both the student and the primary instructor.

Yes, the student must have sufficient knowledge and acceptable performance for the stage check, but there is always room for improvement. The primary instructor conveys a tremendous amount of information to the student, and it's only natural that the more important points get diluted with the not-so-important points. A good stage check instructor will restore the proper emphasis on the important points.

I've been in the general aviation training business for many years, and believe me, what I have explained is a win-win situation for everyone—the student, the instructor, the school, and the industry. I would tell flight school managers who do not operate in this manner that the FAA is not your mother. You must get with the program, show some initiative, standardize your instructors, and structure your program properly.

Yes, it costs money, but quality always does, and a CFI's existence depends on his or her reputation. Why would anyone compromise that?

Our country has produced excellent pilots for years. Today, however, I encounter more airplane drivers than pilots. Drivers fly with their heads buried in the cockpit, primed with rote answers for anticipated questions. Pilots fly with their heads outside the cockpit, primed with knowledge and understanding—they were taught *the art of flying.*

Flight Training Facts

Don't be snookered - September 2003 issue

Ethics in the flight training community have always varied. Quoting minimum flight training times and costs in order to snag a student, continuing to train an unsatisfactory student who is a potential aircraft sale, or flight-time acquisition for an instructor have frequently compromised ethical standards. The unsuspecting student may pay for these indiscretions financially—or with his life or the lives of others.

What should a prospective student do? Do not make assumptions based on first impressions. You must determine the credibility of the person to whom you are speaking.

All flight instruction is not equal, and this has nothing to do with an instructor's age, gender, or experience. Earning a flight instructor certificate requires hard work, and I have known many new instructors who trained students just as well as any experienced instructor. There is a limit as to what a student can learn, and no instructor can teach experience. Experience can only simplify what is taught.

I've never met an insincere flight instructor, but I have met many who are misinformed. That's not their fault, it's just a reflection of the training they received.

You get what you pay for, and you must pay for quality instruction. The cheapest offer is seldom if ever the best deal in the long run.

Good flight instructors are teachers. They don't just point out errors; they tell you why those errors are occurring. They conduct preflight briefings to ensure that you have the proper knowledge before you start a lesson, and they conduct postflight briefings to review what you learned in the air and what to expect on the next lesson.

A good instructor will not advance a student in the training program until the current task has been mastered. Hoping that a current student problem can be solved by moving on only compounds that problem and makes further learning more difficult if not impossible.

Don't assume that your desire to fly qualifies you to fly. An ethical flight instructor will make that determination during the first stage of instruction.

As a student, you have many obligations. You must obtain FAA medical certification, and some medical conditions are disqualifying. Do this before you start training.

Psychological testing is not required, but flight instructors are trained to recognize five hazardous characteristics that are disqualifying unless corrected: *Anti-authority*—willfully violates rules. *Impulsivity*—does not think things through before reacting. *Invulnerability*—assumes that nothing bad will ever happen. *Macho*—takes foolish chances. *Resignation*—gives up when things are not going well.

You must have a good command of the English language, good judgment and awareness, and the ability to constantly divide your attention. You must have the time and desire to study. For every hour that you fly during training, you must crack the books for three to four hours.

You must be prepared for each lesson. Work out a schedule with your instructor that gives you enough study time.

Never fly if one of the *I'm Safe* checklist factors exist: *Illness*—Any symptoms? *Medication*—Taking prescription or over-the-counter drugs? *Stress*—Worried about your job, finances, health problems, or family discord? *Alcohol*—Drinking within eight hours? *Fatigue*—Tired or not adequately rested? *Eating*—Adequately nourished? And, Paragraph 8-1-1 of the *Aeronautical Information Manual* notes, the emotions of anger, depression, and anxiety from such events not only decrease alertness but also may lead to taking risks that border on self-destruction.

Flight training is a team effort. Instructors and students alike must meet their responsibilities. If either one fails to do this, the other person must take action. A responsible student should find another instructor; a responsible instructor should terminate the student.

AIRPLANES

If the proper attitude references and power settings are taught, you will be the master all airplanes.

The type of airplane used for flight instruction is immaterial, a point that was implied in the introduction when reference was made to one of man's great inventions—the wing. If your instructor knows how to teach this common denominator along with the proper attitude references and power settings, you will be the master of all airplanes.

Each type of airplane has unique features and each pilot has personal preferences. Regardless of rumors, it is erroneous to say that one airplane is better than another. It depends on what you like—Ford or Chevy. If a choice of airplanes exist, sit in each one and decide for yourself, but remember, simple airplanes are best for initial training.

Airplane rental charges are based on wet or dry rates and tachometer or Hobbs meter readings. Wet rates include fuel, dry rates do not.

The tachometer is a mechanical device that records engine rpm and hours of operation. This reference is called *tach time.* Being mechanical, it is accurate only when the engine is operating near cruise power. At idle power, it hardly moves. Airplane maintenance inspections are based on *tach time* or time in service—airplane time when airborne.

Rental charges are based on Hobbs time.

The Hobbs meter is an electrical instrument that records hours of operation. This reference, called *hobbs time,* always gives an accurate indication of engine operating time. Most flight schools and flying clubs base rental charges on *hobbs time*, and that time is logged in your personal log book and the flight school's training records.

TEENAGERS AND FLYING

Soloing is a stand-alone goal.

Teenagers who are sixteen or older and want to fly should be given the opportunity to solo an airplane. This rewarding experience gives them additional motivation and confidence in their everyday activities. Soloing does not mandate further flight training. It is a stand-alone goal.

If drugs are encountered, a teenager who has soloed can counteract peer pressure by stating that he is a student pilot who is required to take FAA medical examinations that can detect drug use. His stature is now elevated, and drug participation is easily avoided.

If you desire a flying career, have an alternate career option.

If the student desires a flying career, proceed with caution and plan at least one alternate career option for protection. As I mentioned earlier, the ability to obtain and maintain a good flying job is very unpredictable.

PILOT CERTIFICATES AND RATINGS

The certificates and ratings required depend on your personal objectives. A certificate dictates allowable privileges. Ratings, listed on a certificate, dictate additional privileges.

Sport Pilot Certificate
You can fly light-sport airplanes during the day in visual conditions. You can carry only one passenger. FAA medical certification is not required, but you must possess a valid drivers license. Light-sport airplanes are two-place aircraft with a gross weight of not more than 1,320 pounds.

Student Pilot Certificate
You can fly solo in order to fulfill recreational or private pilot requirements. With the proper flight instructor endorsement, the certificate is valid for 90 days.

Recreational Pilot Certificate
You can fly airplanes, but there are restrictions on aircraft size, passengers carried, cross-country flying, night flying, and operating in airspace that requires ATC communications.

Private Pilot Certificate
You can fly most single-engine airplanes. The limitations that apply to sport and recreational pilots do not apply.

Instrument Rating
You can fly in clouds and in low visibility conditions.

Commercial Certificate
You can receive compensation for your service, but there are limitations on carrying passengers and freight.

Multiengine Rating
You can fly multiengine airplanes.

Flight Instructor – Basic Certificate (CFI)
You can instruct other pilots in single-engine airplanes.

Flight Instructor – Instrument Rating (CFII)
You can instruct other pilots for instrument flying.

Flight Instructor – Multiengine Rating (CFIME)
You can instruct other pilots in multiengine airplanes.

Airline Transport Pilot Certificate (ATP)
You can be a captain for a commercial airline.

"GIVE ME A SPARK OF NATURE'S FIRE,

THAT'S THE LEARNING I DESIRE."

Robert Burns

CHAPTER 2

STANDARD OPERATING PROCEDURES

Standard operating procedures (SOPs) are formalized rules that deal with safety, awareness, and professionalism. When followed, you maximize safety, extend airplane and engine life, and gain the respect of other pilots, mechanics, air traffic controllers, and linemen.

Some pilots fail to consider how others perceive their behavior and actions. Are they calm, efficient, congenial, and organized? Is the airplane operated safely, smoothly, and economically? SOPs are the foundation for these factors. A pilot desiring advancement or acceptance never knows what an observer passes on to others. If you are all thumbs or the airplane is abused, professional advancement and peer acceptance are highly unlikely.

In essence, SOPs standardize all pilots and instructors. The following SOPs apply to fixed-gear, fixed-pitch propeller airplanes. Make these SOPs habitual, but use good judgment—they may not apply to every airplane and every situation.

THE PILOT'S OPERATING HANDBOOK IS THE FINAL AUTHORITY FOR AIRCRAFT OPERATION.

ENGINE OPERATION

GENERAL

Always assume that the magnetos are on.

- Regardless of airplane status, always treat the engine and propeller as though the magnetos are on.

- Study the airplane's pilot's operating handbook (POH) for correct operating procedures. Never base your operating technique on hearsay or personal assumptions.

PREFLIGHT

- If the engine cowl can be opened, check for oil or fuel leaks and the general condition of all items. Most oil, fuel, and pneumatic lines have color coded markings for identification.

For proper cooling, the engine compartment air baffles must be in good condition.

Check the condition of the flexible, rubberized air baffles that are attached to the top of the metal air baffles that surround the engine. These devices trap incoming air and create a high pressure area above the cylinders. That air then flows past the cylinder head cooling fins and escapes into the lower engine cowl. If the baffles are damaged or missing, the high pressure area will not develop, airflow past the cylinders will be reduced, and overheating will occur.

- Most oil dipsticks are round, because oil will climb up a flat dipstick while the engine is shut down. Always wipe the oil off a flat dipstick and reinsert it to check oil quantity. Never let any oil drip into the engine compartment. One or two drops will make a mess due to the airflow that occurs in flight.

When moving the propeller by hand, rotate it backward.

- To inspect the propeller, check compression, or loosen congealed oil in cold weather, move the propeller backward from its normal direction of rotation so that the left magneto's impulse coupling does not activate.

- When rotating the propeller backward, note the resistance as each piston completes its compression stroke. Light resistance indicates weak compression due to worn or damaged piston rings, sticking valves, or burnt exhaust valves. Strong resistance indicates good compression. Extreme resistance indicates sticking valves.

STARTING

Never pump the throttle prior to engine rotation unless authorized by the POH.

• Prime a carbureted engine with the fuel primer. Never pump the throttle to prime an engine unless the starter motor is engaged and rotating the engine. This ensures that the fuel-air mixture will be drawn into the engine.

When you pump the throttle, the carburetor's accelerator pump sprays fuel into the carburetor. Automobiles have downdraft carburetors, so pumping the throttle is acceptable prior to engine rotation because the fuel-air mixture falls into the engine's intake manifold. Most carbureted airplanes, however, have updraft carburetors, and the fuel-air mixture will fall down into the air intake ducting and cause a serious fire hazard if the engine backfires while starting.

Some airplanes have sidedraft carburetors located at the rear of the engine. In this case, the airplane's POH may tell you to pump the throttle a few times before you engage the starter.

• Prolonged engine starts occur because of improper priming or throttle position. The exact amount of prime depends on experience and ambient conditions. Excessive priming causes fuel accumulation in the cylinders that dilutes the oil coating on the cylinder walls and increases engine wear.

If the throttle is open too far (too much air), the mixture will be too lean to start; if closed too far (insufficient air), the mixture will be too rich to start. If you pump the throttle to prime a rotating engine, quickly return the throttle to the start position (1/4 to 1/2 inch open) so that the engine will start.

If an engine does not start within 30 seconds, let the starter motor cool for two minutes.

• Prolonged cranking will damage the starter motor. If the engine does not start within 30 seconds, let the starter motor cool for two minutes before you attempt another start.

• At high density altitudes, the engine will neither start nor idle if the mixture is full rich.

• After a cold-engine start, improper fuel vaporization may cause a rough idle. Increased power may smooth out the idle, but taxiing at that power setting may force you to drag the brakes in order to control taxi speed which is not good.

After engine start in extremely cold weather, use the primer to keep the engine running.

• In extremely cold weather, you may need to use considerable prime and pull the engine through by hand several times before attempting an engine start. After priming, recharge the primer but leave it out. If the engine starts to die, keep it running by supplying fuel with the primer. Do not open the throttle, because too much air will enter the engine and it will stop running. Continue to use the primer until the engine warms up and idles properly. Always close and lock the primer before you perform the engine runup.

TAXIING

• When taxiing, keep the mixture leaned to prevent spark plug fouling, and use power to control taxi speed. Idle power is perfectly acceptable and is frequently required.

• When stopped, use the manufacturer's recommended minimum rpm, usually 1,000 rpm. By observing this limit, enough air passes over the bottom lip of the engine cowl or the cowl flaps to create a low pressure area that helps pull air through the front air inlets and past the cylinders. When taxiing, the relative wind provides sufficient cooling air and you can use idle power.

RUNUP AND TAKEOFF

• The POH may specify a minimum oil temperature for engine runup and takeoff. If one is not specified, takeoff can be made when the engine will accelerate smoothly, but use caution in extremely cold climates. If the oil is not warm, oil pump cavitation and engine failure may occur during takeoff.

Always make smooth power changes.

• Always make smooth power changes. Rapid throttle movement that causes the engine to hesitate or stumble places high stress on the propeller and crankshaft.

• Before making a high altitude runup and takeoff, lean the mixture for best power. When you set runup power or set full power for takeoff, lean for maximum rpm.

• After you set takeoff power, check oil pressure and the rpm to make certain that the engine is developing full power.

Most engine failures occur during the initial power reduction after takeoff.

• After takeoff, never set climb power until you reach a safe altitude—the altitude that gives you a forced-landing option. In the mid-1960s, the U.S. Army stated that most engine failures occur during the initial power reduction after takeoff.

• You are not "saving" an engine if you make an early power reduction after takeoff. This fact is documented by two engine manufacturers—Continental and Lycoming.

CLIMB

Climb to the safe altitude at the best-rate climb speed.

• Climb to the safe altitude using the best-rate climb speed. A slower climb speed offers no advantage unless obstacle clearance is required. A faster climb speed increases engine rpm and propeller noise that will tempt you to make a premature power reduction.

• Upon reaching the safe altitude, establish the cruise-climb attitude and airspeed, and then set climb power.

• A reduced power climb may cause overheating and engine damage. Use the climb power setting specified in the POH. If full throttle is specified, that throttle position is mandatory in order to provide extra fuel for valve cooling.

When power is 75 percent or less, lean the mixture.

• Lean at any altitude if power output is 75 percent or less.

CRUISE

• With no EGT gauge, lean the mixture until the engine runs rough. Now, advance the mixture twice the distance that is required to make the engine run smooth. For example, if you advanced the mixture 1/4 inch to obtain smooth operation, advance it another 1/4 inch for best power. With an EGT gauge, use the POH's leaning procedure.

Always use full carburetor heat unless you have a carburetor air temperature gauge.

• Always use full carburetor heat prior to closing the throttle and when ice is suspected due to an rpm drop. Carburetor ice can occur in clear air with temperatures below 80 degrees Fahrenheit (27 degrees Celsius) and humidity above 50 percent. Use partial heat only if a carburetor air temperature gauge is installed.

Carbureted Lycoming engines are less susceptible to carburetor ice than Continental engines because Lycoming induction manifolds are routed through the engine crankcase. Heat from the engine oil increases the induction air temperature.

If carburetor heat is not removing the ice, apply full power for maximum heat. Caution! Throttle movement in an extreme case may dislodge ice that will block the airflow past the throttle valve and stop the engine. If the engine starts losing power, aggressive leaning may be required in order to keep it running. If continuous, full carburetor heat is required due to icing conditions and the engine runs rough, lean the mixture for smooth engine operation.

DESCENT

- Keep the engine leaned during descents. Richen the mixture just enough to keep the engine running smoothly.

During power-cff descents, clear the engine once every minute.

- When power-off descents are made during training flights and carburetor icing conditions exist, apply full carburetor heat before you reduce power. Momentarily increase power once every minute in order to clear the engine.

- Do not use idle power in the air—particularly with fuel injected engines—unless you performed a satisfactory idle check after the engine run-up.

SHUTDOWN

A broken P-lead is a critical safety hazard.

- Check the magneto ground wires (P-leads). Set idle rpm, and then quickly turn the magneto switch off and back on. If the engine did not stop running when the switch was off, a ground wire is broken and that magneto will not be deactivated when the engine is shutdown. You must place a warning sign on the propeller and advise maintenance. If an unsuspecting person moved the propeller, the engine could kick over even though the mixture control is at idle cutoff.

- If you are located at a high altitude airport, note the leanest mixture setting that gave you a smooth idle before engine shutdown. That mixture setting will be required for the subsequent engine start, not full rich.

- Set 1,000 rpm and shut the engine down by moving the mixture control to the idle cutoff position.

You stop the engine with the mixture control.

- Always stop the engine with the mixture control so that raw fuel is not left in the cylinders. This may prevent an inadvertent engine start-up and accident if a magneto was not grounded and an unsuspecting person moved the propeller.

- Turn the magnetos off, and place the ignition key on top of the glareshield. This is a safety procedure so that approaching pilots and linemen can verify magneto status.

IRREGULARITIES OR EMERGENCIES

- If an engine quits running due to fuel starvation, retard the throttle to prevent a power surge when you restore fuel flow.

- If an engine starts to run rough and mixture is not the problem, select each magneto individually. If one is bad, select the good magneto and land as soon as practical.

- If the alternator fails, do not reset it until you turn off all avionics equipment in order avoid damage from voltage spikes that may occur when the alternator restarts. You must memorize the alternator reset procedure that is listed in the airplane's POH.

Turning off the master switch does not affect the engine's magneto system.

To maintain battery charge if the alternator has failed, turn off the master switch if conditions allow, but don't wait until the battery becomes weak. If voltage is too low, the battery contactor will not close when the master switch is turned back on, and the electrical system will remain dead.

CHECKLISTS

- Pilots who do not use written checklists make mistakes.

- Pilots who use only written checklists make mistakes.

Use flow patterns, mental checklists, *and* written checklists.

- Competent pilots use flow patterns, mental checklists, *and* written checklists to prevent error. This provides the necessary redundancy that is mandatory if you want to avoid the shortcomings of human behavior.

FLOW PATTERNS

A flow pattern is a logical path that you follow in order to configure the cockpit for a desired task; pilots who use flow patterns develop a high level of cockpit familiarity that is a real advantage when a timely assessment of cockpit status is needed. Developing a good flow pattern is easy when you realize that all cockpit layouts have a certain degree of logic.

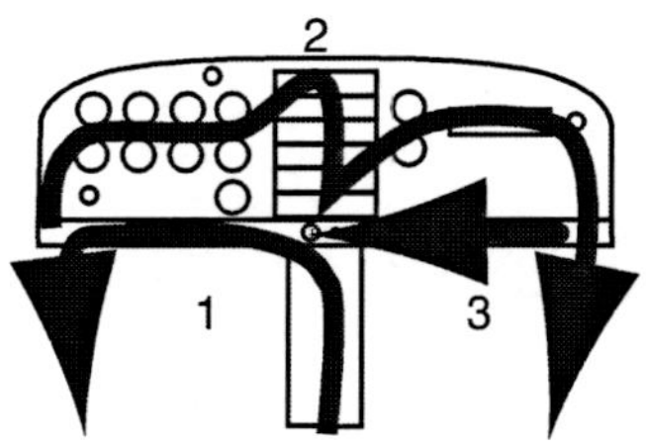

A Cockpit Flow Pattern

Follow the illustration on the left. FIrst, you start on the floor between the pilot seats, move forward along the center console, go up to the throttle, and move to the left sidewall via the lower portion of the left instrument panel. Second, you move from left to right along the entire upper instrument panel and check the right sidewall, and third, you return to the throttle via the lower portion of the right instrument panel.

MENTAL CHECKLISTS

These unwritten checklists are used to remember routine procedures that occur at specific points on every flight. CIGARS; GUMPS; and Lights-Camera-Action are the mnemonics that recall the most popular mental checklists.

CFI Notes
Teaching the mental checklists is optional during presolo instruction, but mandatory after that. Explain that some items are not applicable to fixed-gear trainers, but establishing those habit patterns—such as routinely saying GUMPS—prepares students for the future when they will probably transition to complex, high-performance airplanes.

CIGARS
Perform this mental checklist before you use the written before-takeoff checklist. If you are well clear of ramp congestion and other airplanes, accomplish the first four items (CIGA) while taxiing to the run-up area. The instruments that are prefaced with an asterisk must be checked while taxiing.

C - Controls: Free with correct movement. If you extend the thumb of the hand that is holding the yoke, it will point to the up-aileron if the ailerons are rigged correctly.

* The four instruments that must be checked while taxiing.

I - Instruments: Correct indications.
Airspeed indicator—zero.
Altimeter—indicates within 75 feet of field elevation when set to the current altimeter setting.
Vertical speed indicator—zero.
*Magnetic compass—turns freely, agrees with the runway heading, and the compass fluid level is not visible.
*Attitude indicator—does not bank in excess of five degrees when turning.
*Heading indicator—turns freely and is aligned with the magnetic compass.
*Turn coordinator—indicates a turn, and the ball moves to the outside of that turn.

G - Gas: Fuel on proper tank with boost pump off to ensure operation of the engine-driven fuel pump.

A - Attitude Trim: Recheck all trim controls for takeoff.

R - Run-up and Radios: (Discussed in the ground operations section of this chapter.)

S - Safety Belts, Doors and Windows: Fastened and closed.

GUMPS

You should complete the written before-landing checklist—with the exception of the GUMPS items—just before you start your descent from cruise altitude. Now, perform the GUMPS mental checklist three times: when you are cleared to land, when you leave or pass through pattern altitude, and when you turn onto the final approach. In some cases you'll only be able to do this twice, but at least you'll never forget to lower the landing gear if you make this procedure habitual.

G - Gas: On the fullest tank with fuel pump on if required.

U - Undercarriage: Landing gear down and locked.

M - Mixture: Set for landing.

P - Propeller: High rpm position.

S - Safety Belts: Fastened.

LIGHTS-CAMERA-ACTION

Perform this mental checklist when you taxi onto the runway for takeoff, after you reach cruise altitude, before you leave cruise altitude, and when you clear the runway after landing.

Taxiing onto the runway

Lights - Exterior lights as required.
Camera - Transponder on.
Action - Time, oil pressure and takeoff rpm checked.

Reaching cruise altitude

Lights - Exterior lights as required.
Camera - Not applicable.
Action - Cowl flaps closed, fuel pump off (if not done earlier), power and mixture set, engine instruments and circuit breakers checked, heading indicator aligned with the magnetic compass.

Leaving cruise altitude

Lights - Exterior lights as required.
Camera - Not applicable.
Action - Heading indicator aligned with magnetic compass, altimeter set, circuit breakers checked, written before-landing checklist completed except for the GUMPS items.

Clear of the runway

Lights - Exterior lights as required.
Camera - Transponder on standby.
Action - Call for taxi clearance first so that you don't hold up traffic, and then use a flow pattern to clean up the airplane: cowl flaps open, wing flaps up, mixture leaned, carburetor heat and fuel pump off.

WRITTEN CHECKLISTS

The written checklist is the final authority for operating the airplane.

These checklists are the final authority for aircraft operation, and they must be used in conjunction with flow patterns and mental checklists as the following examples describe. The written checklists are a checklist, not a do-list. You are checking previous actions to ensure compliance. Student pilots must always announce to their instructor or the examiner that the written checklists denoted with an asterisk in the following examples have been completed.

*** Preflight Inspection**
You will learn to use a flow pattern for conducting the preflight inspection. Upon completion, review the written preflight checklist to ensure compliance.

*** Before Engine Start**
Use a flow pattern to prepare the cockpit for engine start. Before you yell "clear" and start the engine, review the written before-start checklist to ensure compliance.

*** Before Takeoff**
While taxiing to the run-up area, start the CIGARS mental checklist. After you complete CIGARS and use the flow pattern to prepare the airplane for takeoff, review the written before-takeoff checklist to ensure compliance. This checklist—which includes flaps, mixture, fuel pump (if required), doors and windows—must be completed *before* you call the tower for takeoff clearance.

* You must announce that these written checklists have been completed:

Preflight
Before Engine Start
Before Takeoff
Before Landing
Engine Shutdown

Taxiing Onto the Runway
Execute the lights-camera-action mental checklist.

After You Reach Cruise Altitude
Execute the lights-camera-action mental checklist.

*** Before You Leave Cruise Altitude**
Execute the lights-camera-action mental checklist, and with the exception of the GUMPS items, complete the written before-landing checklist.

When Cleared to Land, When Leaving Pattern Altitude, *and* When On Final Approach
Execute the GUMPS mental checklist at each point.

After Landing
When stopped clear of the runway, call for taxi clearance and then execute the lights-camera-action mental checklist. Never retract the flaps, move switches or system controls, or call ground control while you are on the runway.

*** Engine Shutdown**
Use the flow pattern to shut down the airplane, and then use the written engine-shutdown checklist to ensure compliance.

Irregular or Emergency Situations
Use a flow pattern and good judgment to initially assess the situation. Take corrective action when a solution is obvious. Then, if time permits, use the appropriate written checklist to ensure compliance.

An emergency requires both immediate and reference actions; an irregularity requires timely reference actions. Immediate action items are performed from memory. Reference action items are performed using a written checklist.

COMMUNICATIONS AND AVIONICS

- Protect your battery; minimize electrical loads when the engine is not operating. It's best to start the engine before you use avionics equipment. If necessary for a brief call, you can use one communication radio prior to engine start *providing* the remaining avionics equipment is turned off.

Listen before you transmit; plan your transmission before you talk.

- Listen before you transmit; think before you talk. In other words, do not interrupt communications that are in progress, and plan your transmission before you key the mike.

- Check the transmitter selector position and the receiver volume before you key the mike. To set the receiver's volume, turn the squelch control to the maximum sensitivity position, adjust the volume so that the squelch noise is at a comfortable level, and then turn the squelch control down until the noise just disappears. With the squelch control in this position, receiver sensitivity is maximized.

- When inbound to an airport contact the appropriate facility at the recommended distance: 25 miles for approach control, 15 miles for control towers, 10 miles for unicom stations or local airport advisory service (LAA) from a flight service station (FSS).

Make accurate position reports

- Make accurate position reports so that other pilots on that frequency can find you if they are nearby. If you can't report over an intended point due to frequency congestion, report in reference to that point. For example, Skyroamer 17-Romeo, two miles south of the dam, 2,500, landing with X-ray.

- Communicate numbers individually: "Skyroamer 1-7-Romeo," "1-5 miles south," "2 thousand 5 hundred," "Runway 2-6-left.".

Use your abbreviated call sign only after a controller initiates it.

- Use your full call sign on the initial call to a facility—Skyroamer 9117-Romeo. If the controller acknowledges your call using an abbreviated call sign—Skyroamer 17-Romeo—you should do likewise on subsequent transmissions. Do not say, "17-Romeo." Always state the aircraft prefix—Skyroamer, Cessna, Piper, etc.—and then the registration number.

If no response is obtained on the first call, wait a few seconds before calling again. The controller may be busy coordinating traffic with another controller or may be talking on another frequency.

- You must always acknowledge a controller's clearance or instruction. However, the only items that you must read back are taxi, hold short, runway crossing, and position and hold instructions; and takeoff and landing clearances. Example: "Skyroamer 17-Romeo, taxi to Runway 26-left via Bravo and Lima. Hold short of Runway 26-right." You respond, "Skyroamer 17-Romeo, 26-left, Bravo, Lima, hold short 26-right."

Do not read back non-critical instructions or advisories that you understand.

For other instructions or advisories, use the word *wilco* (will comply) or *roger* (I understand) to acknowledge. However, if you're not certain that you heard the transmission correctly, do read it back. To read back those that you do understand creates terrible frequency congestion that is very annoying to pilots and controllers alike. Examples: "Skyroamer 17-Romeo, follow the Baron on base leg" or "Skyroamer 17-Romeo, give way to the United 737 and taxi to parking." In both cases just say, "Skyroamer 17-Romeo, Wilco."

Caution! Never acknowledge a radio transmission by double clicking the mike. The controller has no idea if the transmission was received by the intended aircraft.

The radio transmission rule: who, you, where, what.

- When you initiate a call, four elements must be transmitted: Who—the station you're calling. You—your identification. Where—your location to include altitude when airborne. What—your request. Here are some examples:

Taxi For Takeoff

"Chino Ground, Skyroamer 9117-Romeo, northwest tiedowns, taxi for takeoff with X-ray."

"Big Bear Unicom, Skyroamer 9117-Romeo, transit ramp, taxiing to Runway 8 for takeoff."

Takeoff

"Santa Monica Tower, Skyroamer 9117-Romeo, ready for takeoff." Make this call when you are number one at the approach end of the active runway and you have completed the written before-takeoff checklist. State the runway number if multiple runways are in use.

"Chino Tower, Skyroamer 9117-Romeo, ready for takeoff, 26-left, request closed traffic."

"Chino Tower, Skyroamer 9117-Romeo, ready for takeoff, 26-left, request downwind departure."

Landing

"Chino Tower, Skyroamer 9117-Romeo, 15 south, 2,500, landing with X-ray."

Taxi After Landing

"Chino Ground, Skyroamer 9117-Romeo, clear of 26-right at Bravo, taxi to northwest tie-downs."

- Be concise and explicit. Controllers will attempt to honor any request if you transmit it early and clearly.

For example, you are south of an airport that has parallel east-west runways and you want to enter the north traffic pattern for touch-and-go landings on Runway 26-right. Give the tower this basic request, not your desired plan of action.

Correct: Chino Tower, Skyroamer 9117-Romeo, 15 south with X-ray, 2,500 feet, request closed traffic on 26-right.

Incorrect: Chino Tower, Skyroamer 9117-Romeo, 15 south with X-ray, 2,500, request an extended left base entry for closed traffic on Runway 26-right.

Listen and keep an open mind. Do not anticipate what will occur, because controllers have several options that depend on current traffic conditions. You may be cleared to overfly the airport and enter the north pattern, to fly an extended base leg entry to Runway 26-right, or to make a touch and go landing on Runway 26-left and then enter the north pattern.

• If you are using a microphone instead of a headset with a boom mike, stow the microphone properly when you finish transmitting in order to prevent accidental activation of the transmit button.

• When asked, "How do you hear this transmitter?" respond in terms of loudness and clarity. For example, "loud and clear," "weak but clear," or "loud but garbled." Years ago, a numerical scale from 1 to 5 was used for loudness and clarity. "5 by 5" meant loud and clear.

• Always operate the transponder on standby when taxiing, and turn it on or off in conjunction with the lights-camera-action mental checklist. When changing codes, do not accidentally place 75, 76, or 77 in the first two digits. Regardless of the last two digits, 7500, 7600, or 7700 will be transmitted.

Avoid voltage spikes. Turn the radios off before the alternator is turned on or off.

• To prevent voltage spikes from damaging avionics equipment, turn that equipment off before you turn the alternator on or off. That includes engine start and shutdown.

GROUND OPERATIONS

GENERAL

• Be alert. Watch! Listen! Think! Anticipate! Act!

• Know the location of fire alarms and fire extinguishers at home base and at frequently used aviation facilities.

Keep smokers away from parked aircraft.

• Never allow smoking within 100 feet of aircraft, fuel trucks, or fuel facilities, or in an airplane that is not airborne. Depending on fuel loads and atmospheric conditions, fuel tank vents of parked or taxiing aircraft will emit fuel vapor and liquid fuel.

• Do not smoke in an airplane during ground operations. If you allow smoking when airborne, use extreme caution. You never want to experience an inflight cabin fire.

• If you need to get the attention of a pilot in an aircraft that has its engine or engines running, approach an airplane from the left side but remain clear of propellers or turbine inlets; approach a helicopter from the front but remain clear of the area under the main rotor blades.

• Keep all bystanders clear of airplane propellers, jet intakes and exhausts, and helicopter rotor blades.

• Stay well clear of large airplanes and helicopters unless the location of the jet blast, prop wash, or rotor wash is known and avoidable.

• Always be acutely aware of the actual wind conditions and the airport's ground and airborne traffic flow.

AIRCRAFT SERVICING

Accidents have occurred because aircraft were serviced with the wrong fuel.

• Remove all occupants prior to refueling.

• Check the grade of fuel and oil that is being supplied.

• Remain with the airplane to ensure proper servicing.

• Ground the fuel truck to the airplane before refueling. Otherwise, a difference in electrical potential between the airplane's structure and the fuel nozzle could generate a spark and start a fire. The airplane connection point for the fuel truck's grounding wire must be an unpainted surface other than the exhaust stack. The fuel truck must also be grounded to Earth using a ramp tie-down anchor.

For a full fuel load, you may need a specific fuel selector position.

• Consult the POH for refueling procedures. If the airplane has interconnected fuel tanks, a specific fuel selector position may be required in order to completely fill all tanks.

When cleaning plexiglas, use linear motion, not circular motion.

• Plexiglas (acrylic) windows can be kept free of scratches if dust and dirt are removed with a bare hand while the windows are flushed with water. Move your hand up and down in a straight line between the top and bottom of each window. Circular motion causes minor scratches to become more apparent. Dry with a damp chamois to prevent static electricity buildup that attracts dust.

To remove stubborn dried-on bug residue, use a 50/50 mixture of warm water and isopropyl alcohol. Finally, buff the windows with a soft towel and plastic polish.

• To remove exhaust stains from the fuselage, many POHs tell you to use Stoddard solvent, an old name for Naptha or lighter fluid.

• Keep grease (includes lipstick and mustache wax) away from oxygen fittings. When pure oxygen is present, grease can spontaneously ignite.

• You can measure the fuel quantity in a partially filled tank if you make a calibrated dip stick for that tank. Fuel capacity can change with attitude changes, so park on level ground with the airplane unloaded in order to establish a reference attitude. Then, add fuel to the previously drained tank in five gallon increments, marking the dip stick at each interval.

• Always fill the fuel tanks if an airplane is to sit for any appreciable time. Full tanks prevent water condensation and keep rubber fuel cells from drying out.

Never trust published fuel tank capacities.

• Never trust published fuel capacity unless you personally know how much usable fuel the tank will hold. A full tank does not mean it contains what is stated in the POH. Fuel tanks can partially collapse, and some tanks cannot be completely filled unless the airplane is in a specific attitude.

• Always recheck fuel and oil quantities and caps after servicing. Good linemen expect this from a safe pilot.

• When an airplane will be overweight, you must remove fuel until you are at the gross weight limit. In years past, a fuel truck could do this, but today most airports require fuelers to pump the fuel out by hand and then have it refiltered.

PREFLIGHT

If you hoot with the owls, you can't soar with the eagles.

• Before leaving home, preflight yourself. Are you fit to fly? Never fly when you are ill, taking medication, under stress, under the influence of alcohol or drugs, fatigued, or emotionally distraught.

• Before you depart on a flight, you must check the following items: airport departure procedures, weather, winds aloft, notams, TFRs, and the airplane's airworthiness. Airworthiness implies a thorough preflight inspection and the required documentation and maintenance inspections.

• While walking out to the airplane note the wind condition, airport traffic flow, congestion around the airplane, and the best way to exit the tie-down area.

• At the airplane, check the serviceable items first—windshield cleanliness, fuel, oil, hydraulic fluid, and tire inflation. If service is required, a call at this time can prevent a delay.

• Use caution when opening cabin doors. Wind or another airplane's prop wash can snatch a door out of your hand, slam it open, and cause expensive damage.

When the engine is shutdown, always leave the ignition key on top of the glareshield.

• After unlocking the airplane, place the ignition key on top of the glare shield until you are ready for engine start. This safety precaution prevents passengers from activating the starter and it indicates ignition status to linemen.

• Perform a thorough initial cockpit check before preflighting the airplane. Here is a typical sequence:

The power-off check "AROW" check (**A**irworthiness certificate, **R**egistration, **O**perating limitations (found in the POH and on aircraft placards), and **W**eight and balance data), Hobbs meter reading, control lock removed, electrical switches off, throttle idle, mixture at idle cutoff, trim controls neutral, fuel selector on, cowl flaps open, and flaps extended if operated manually.

The power-on check Master switch on, fuel quantity, alternator warning light, the exterior lights and stall warning horn (if electric), and flaps extended if operated electrically.

Check the alternator warning light before engine start. Otherwise the alternator must be turned off and on with the engine running, an action that may cause voltage spikes and damage to operating avionics equipment. If the light has a push to test button, only the bulb is tested, not the system.

Fire and midair collisions—two things that pilots must fear.

• There are only two things that pilots must fear when flying: fire and midair collisions. Do not fly unless the airplane has a fire extinguisher within reach of your seat.

• Radios, headsets, microphones, and cabin speakers do fail. One option is to carry a hand-held aviation transceiver.

Keep loose objects off the cockpit floor.

• Keep loose objects that could jam a rudder pedal off the floor. Store laminated checklists, manuals, and personal equipment in sidewall pockets or between the seats.

• Perform the preflight inspection in accordance with the POH. The following items supplement that material.

• If anyone is inside the airplane's cabin, remain clear of the propeller.

• In addition to the propeller's leading edge, inspect the face of each blade (the surface that faces the cockpit) for nicks or gouges and inspect the propeller spinner for cracks. The back of each the blade (the surface that faces forward) rarely becomes damaged. The propeller spinner is required equipment on most airplanes, because it helps to direct airflow so that the engine will not overheat.

• If the propeller must be moved, rotate it backwards to prevent the magneto's impulse coupling from engaging.

Never push on the nose spinner, nose cowl, or tail surfaces.

• Roll the airplane forward or backward to inspect the tires for bald spots and valve stem alignment. If a tire is underinflated, the tube can slip on the rim and place high stress on the valve stem. To push or pull using the propeller, place both hands close to and on opposite sides of the spinner. Never push on the nose spinner, nose cowl, or tail surfaces.

• All fuel tank vents must be clear of obstructions. A plugged vent will allow negative pressure to develop in the tank that will stop fuel flow to the engine and may even collapse the tank—the fuel quantity indication will then increase. Fuel cap seals must be pliable and not cracked. Otherwise fuel will siphon out in flight due to the decreased air pressure above the wing, and rain or wash water will enter the tank when the airplane is on the ground.

To sample fuel, always start with the highest drain and work down to the lowest.

• After fueling, drain a fuel sample from each location starting with the highest drain and finishing at the lowest drain. If water exists, it will quickly settle to the lowest point and be clearly visible in the bottom of the sample unless a large amount of water exists. To verify fuel color, look vertically through the sample or hold it against a white background.

If the tanks are rubber fuel cells with flat bottoms, water can be trapped if folds exist at the bottom of the cell. Rocking the wings of these airplanes allows trapped water to flow toward the drains.

When the fuel emitted from a drain cannot be captured in a container, check the pavement immediately after draining. Usually the ground will absorb gasoline, but water, if present, will momentarily pool. This procedure is used only as a last resort.

• Ensure that all airframe moisture drain holes are open. Located on each airframe component, they permit water accumulated from rain, washing, or condensation to escape. Plugged drain holes can trap water that might cause problems, particularly if freezing temperatures occur.

You must remove *all* frost, ice, and snow prior to flight.

• Remove all frost, ice, and snow from the airframe prior to flight. A light coating of frost is enough to severely disrupt the airflow and severely degrade airplane performance. Small amounts of ice or snow can unbalance a flight control and cause control flutter, which can cause structural damage. Ice that has formed inside a propeller spinner while the airplane was tied-down will cause a significant vibration.

Never pour hot water on the airplane in freezing condition, and never use automotive type scrapers or deicing fluid. If proper deicing service or preheating is not available, try using a soft brush or a broom. As a last resort, make a double-wall hot water bag using two, heavy-duty plastic trash sacks. Slide the water bag over the contaminated surfaces.

If the crankcase breather tube is frozen shut during cold weather, the crankcase will pressurize and blow oil out through the filler pipe or the engine oil seals. Engine failure is imminent. Check the tube if you can open the engine cowl.

Wipe oil deposits off the lower radio antennas.

• Check the condition of all radio antennas, and wipe oil deposits off the antennas that are located on the bottom of the fuselage. Know the function of each antenna.

• Check all visible control surface and trim tab push-pull rods for binding or excessive end play.

• Check the condition of the airplane's tail tie-down ring. If it is damaged or jammed into the tail cone, the previous pilot may have overrotated the airplane or made a hard landing. Have a mechanic check for internal fuselage damage.

• Position and secure all baggage in accordance with weight and balance limitations and POH requirements.

• Check that all emergency exits can be opened from inside the cabin. Some doors with keyed, external handles must be left unlocked in order to be used as an emergency exit.

Before takeoff, you must brief passengers on how to fasten and release their safety belts.

• Federal regulations require you to brief your passengers on how to fasten and unfasten their safety belts. You should also mention propeller safety, smoking policy, the operation of cabin doors and emergency exits, and ask passengers to help you spot other airplanes in flight. To the passenger in the copilot's seat, point out the hazards of grabbing the yoke or moving the rudder pedals or seat adjusting mechanism.

• Make the flight's final go/no-go decision—you, the airplane, and the weather.

Never start the engine at a location where prop wash will blow toward adjacent airplanes or an open hangar.

• Use a tow bar to pull the airplane out of the tie-down spot to a position where the prop wash from engine start and initial taxi will not blow toward adjacent airplanes or into an open hangar. If the airplane can be started in the tie-down spot, pull it forward of tie-down anchor cables or other obstruction that to cross would require a high power setting.

AIRCRAFT LIGHTS

• Current regulations create confusion regarding the use of exterior lights because of the various lighting configurations on modern airplanes. The following procedures comply with FAA Advisory Circular 91-73A, published in 2003.

- Check the operation of all exterior lights prior to flight.

Never use the strobe lights during taxi if they will adversely affect the vision of other pilots or ground personnel.

- Aircraft lights may be used to make an aircraft operating on the airport surface at night more conspicuous. Pilots may use various combinations of exterior lights to convey their location and intent to other pilots. Strobe lights should never be on during taxi if they will adversely affect the vision of other pilots or ground personnel.

- Engine running: Turn on the rotating beacon (if installed).

- Turn on all anticollision and landing lights when taxiing onto the runway if cleared for takeoff. If cleared "into position and hold," however, do not turn the landing light on until you receive your takeoff clearance.

- Turn the landing light on when operating in high density airspace or within 10 miles of an airport.

- Leave all lights on when operating in reduced visibility.

STARTING

- Check the surrounding area for people, and call "clear" in a loud voice through an open window or door.

- Fuel flow from all tanks must be verified before takeoff. If both wing tanks are not used simultaneously, select the tank with the least fuel for engine start and initial taxi. Switch to the fullest tank for the engine run-up and the takeoff.

- When the engine starts, check these items using the following sequence:

Minimize propblast during engine start.

RPM - Watch the tachometer and do not exceed 1,000 rpm. This action will minimize propblast.

Oil Pressure - Check for pressure. If no response occurs within 30 seconds, shutdown the engine.

Mixture Control - Lean. This eliminates excess fuel from the cylinders that will cause spark plug fouling as the airplane is taxied to the run-up area.

As the mixture is leaned with a properly adjusted carburetor or fuel injection unit, the rpm will initially remain constant and then increase slightly just before the engine quits running. The point of maximum rpm represents the optimum fuel-air mixture for idle power. The mixture should be set slightly ahead of this point to permit smooth engine operation when power is advanced for taxiing.

Never fly at night with a weak airplane battery.

- Never fly at night if the airplane's battery, your emergency electrical supply, is weak. Battery condition is hard to check in most light airplanes, but engine start gives a good indication. Does the starter readily crank the engine, or does the starter labor? After engine start, is the ammeter charge rate normal or abnormally high? A prolonged, high charge rate indicates a weak battery.

To determine how long you can fly on battery power, add up the current requirement (amperage) for one communication and navigation radio, the turn coordinator, and any other electrical equipment that you consider necessary. Divide that number into the battery's amp-hour rating. The answer will probably be in excess of 30 minutes, but that number is valid only if the battery is in good condition and fully charged.

RUN-UP

- Run-up on a surface that is free from loose sand and pebbles in order to avoid damage to the propeller and the horizontal stabilizer's leading edge. Clear the area behind the airplane before you increase power.

Perform the engine run-up and takeoff on the fullest fuel tank.

- Select the fullest fuel tank for the run-up and takeoff.

- Select the takeoff mixture setting (full rich at sea level). After the engine run-up and the written before-takeoff checklist have been completed, leave the mixture set for takeoff. Note: The written before-takeoff checklist must be completed before you call for takeoff clearance.

- Check oil temperature. Some airplanes have POH limitations that require a specific oil temperature before you conduct either the run-up or the takeoff.

- Advance power to the run-up rpm and use a flow pattern that starts just past and ends at the magneto switch in order to check each of the items listed below. Be thorough, but do not waste time because ram air for engine cooling is minimal.

Ammeter (zero center) **or Loadmeter** (zero left) - Turn on two or more high load items (pitot heat and landing lights) to check the alternator operation. If the airplane has an ammeter, it should momentarily increase and then return to zero; if the airplane has a loadmeter, it should increase. This is a positive check of the alternator and the voltage regulator, the electrical system's brain.

Vacuum - Within limits.

Air used for carburetor heat is unfiltered.

Carburetor Heat - Check for an rpm drop. Leave the heat on long enough to ensure that a subsequent rise does not occur, a rise that would indicate an accumulation of carburetor ice during taxi. Minimize the use of carburetor heat on the ground, because the heated intake air is unfiltered.

Mixture (optional) - Slowly retard the mixture until engine rpm starts to drop and then return to the proper setting. This lets you determine the sensitivity of the mixture control so that an excessive power drop is avoided when the mixture is leaned in flight. It also helps to eliminate spark plug lead deposits by momentarily raising cylinder temperatures.

Magnetos - Check for the proper rpm drop. If there is no drop or if the drop is excessive, a magneto problem exists.

The magnetos are the last item you check prior to the idle check. This allows the fuel's lead-scavenging agents to remove spark plug lead deposits while the engine is operating at run-up rpm prior to the magneto check.

If the magneto switch is accidentally turned off, leave it off, close the throttle, and then turn the switch back on.

Caution! If the magneto switch is accidentally turned off during the magneto check, leave it off, close the throttle, and then turn the switch back on. If the magneto switch is turned on while the engine is at run-up rpm, the resulting backfire can damage the air baffles in the exhaust muffler. That damage will impede the flow of exhaust air and cause a permanent power loss.

There are two schools of thought regarding magneto checks. The first wants to ensure that both magnetos are on for takeoff. By checking the right magneto first (two clicks to the left), the rotary switch will always end up with both magnetos on.

Never turn the magneto switch off during the magneto check.

The second wants to avoid turning the magnetos off during the run-up at high engine rpm. By checking the left magneto first (one click to the left), you develop an awareness of switch sensitivity, and you are less likely to turn both magnetos off when checking the right magneto.

While both cases are valid, the second makes more sense because it avoids backfiring and muffler damage. The first case is backed up by the written before-takeoff checklist.

An idle check is mandatory even though it's not listed on the some manufacturer's checklists.

Idle Check—smoothly close the throttle and make certain the engine will idle properly.

Caution! Don't fly an airplane if the engine will not idle properly. With fuel injected engines, this situation causes landing accidents. A pilot, high on the approach, pulls the throttle to idle unaware that the engine will not idle. Radiated heat then vaporizes the fuel in the fuel distribution lines. When back on the glidepath, the throttle is advanced, but nothing happens. If on short final, there is not enough time to restart the engine, and the airplane crashes short of the runway.

Carbureted engines do not have the fuel vaporization problem, because when the throttle is advanced, low pressure in the carburetor's throat immediately draws fuel from the carburetor's fuel bowl and the engine accelerates.

TAXIING

- Taxi with your heels on the cockpit floor and the balls of your feet on the rudder pedals. Move your feet up on the rudder pedals only when necessary to activate the wheel brakes.

- As the airplane starts to move, smoothly check the brakes. Abrupt brake application causes lurching. In cold weather, ice on the brakes and calipers may cause brake lockup. If water vapor inadvertently enters a brake line during servicing, condenses, and freezes, that brake may not work.

- Know who is taxiing the airplane. When two pilots are flying, use verbal commands whenever control transfer occurs. The pilot taking control states "I have it." The pilot relinquishing control states "You have it."

Practice stopping with one failed brake.

- If one brake fails while taxiing, maintain directional control with the rudder pedals while applying the good brake to stop the airplane. You should practice this important maneuver.

- Position the ailerons and elevator to maintain control and prevent upset during the following conditions: strong winds, propwash, jet exhaust, and rotorwash. Here's the rule: Dive away from a tailwind; climb into a headwind.

- Taxi no faster than a man can run. Evaluate taxi speed by observing the ground as it passes by either wingtip. If you look straight ahead over the nose cowl, taxi speed always appears to be too slow.

Avoid using brakes to control speed while power is applied.

- If taxiing too fast, set idle power to control taxi speed before you use the brakes. You use power and one brake simultaneously only when a minimum-radius turn is required.

Never drag the brakes. Use spot breaking.

- If taxi speed continues to build with power at idle, use spot braking rather than continuous braking in order to minimize brake wear and keep the brakes cool—you may need maximum braking for an aborted takeoff. To spot brake, apply the brakes smoothly and slow the airplane well below normal taxi speed. Then, release the brakes and let the airplane accelerate slightly above normal taxi speed. Repeat this technique as necessary. Spot braking is normally required when taxiing downhill or downwind.

- If flying a tricycle-gear airplane and taxiing over a rough or uneven surface, hold the yoke full aft in order to minimize stress on the nose strut and maintain propeller clearance.

- If flying a tricycle-gear airplane and you are requested to expedite taxi at a busy airport, hold the control yoke full aft in order to minimize the weight on the nose wheel and increase speed slightly above normal taxi speed. Never do this when you must position the controls for wind, jet blast, or prop wash; or when you are taxiing near other aircraft.

• Always watch for wind indicators prior to takeoff. Winds constantly change, and reported winds are historical events.

Always check magnetic compass alignment when taxiing perpendicular to or parallel to a runway.

• Check magnetic compass alignment when taxiing perpendicular to or parallel to a runway. Set the heading indicator to agree with the magnetic compass, but before you remove your hand from the alignment knob, twist it to ensure that the knob has disengaged from the compass card.

• Approaching the run-up area, clear the area that will be behind the airplane during the run-up. If the run-up area is not crowded, park at the end so that the maximum amount of space is available for other airplanes.

• If the run-up area is adjacent to a runway used by large turbojet airplanes, stay well clear of the area where jet exhaust (jet blast) will be directed when the jet's power is increased in order to taxi onto the runway.

If you must be in close proximity to jet airplanes, park in the run-up pad with your nose is pointed toward the area where jet blast might originate. If inadvertently exposed to jet blast, head-on exposure is safest.

If forced to taxi directly behind a jet on a wide taxiway, you can safely do so if you follow two rules: Stay extremely close to the jet because jet exhaust fans-out with distance, but position yourself on the far edge of the taxiway that will always be clear of jet blast—the taxiway edge that will definitely be outboard of the jet engines and on the inside of the jet's next turn. If the jet will turn left, stay on the far left side of the taxiway; if it will turn right, stay on the far right side.

If you appear to be in trouble from potential jet blast, advise the tower immediately. The controller will alert the pilot and give you a chance to reposition.

Never change fuel tanks just prior to takeoff.

• Never change fuel tanks just prior to takeoff. Always depart on the tank that was used during the engine run-up.

• When stopped on a taxiway or on the runway after the engine run-up and prior to takeoff, set at least 1,000 rpm for engine cooling and to help prevent spark plug fouling.

• When located on a taxiway and cleared for takeoff, do not make a high speed, tight turn when taking the runway unless your fuel tanks are full. With partially full tanks it is possible to uncover the fuel inlets and ingest air in the fuel line. Engine stoppage could occur just after lift-off.

• Never look inside the cockpit when the airplane is on the ground and moving faster than normal taxi speed.

Do not perform tail swings in the tie-down area.

• When taxiing into the tie-down area, do not perform high-powered tail-swings in order to align the airplane with the tie-down spot if your propwash will blast bystanders, other airplanes, or open hangars.

POSTFLIGHT

• Check 121.5 MHz to make certain your emergency locator transmitter (ELT) has not been activated. If possible, set the radios for the next pilot—ATIS and ground control.

• When the engine is stopped, remove the keys from the ignition and place them on the glareshield. This is a safety precaution for observant linemen, instructors. or pilots who may offer to push the airplane into the tie-down spot while you steer from the cockpit.

Use a tow bar to move the airplane into the tie-down spot.

• Use a tow bar to move the airplane into the tie-down spot. Push with one hand on the prop—close to the spinner—the other hand on the tow bar. Helpers should push on a wing or a wing strut. Never push on the prop spinner. If the airplane is hard to move, push with both hands on the prop—close to the spinner—and with the tow bar lying on the ground, control it with one foot or the other.

• Conduct a postflight inspection to check for damage or fluid leaks, remove all cabin trash, neutralize the trim selectors, secure the flight controls, record Hobbs meter and tachometer times, make maintenance write-ups if necessary, recheck that the master switch is off, and lock the airplane.

• If required by local fueling procedures, leave the propeller in the vertical position to indicate that fuel is required.

- Leave the parking brake off if the wheels are chocked, and tie-down the airplane as discussed next or as illustrated on the next page.

At tie-down areas where the airplane tie-down chains are fastened to flexible steel cables that lie flat on the ground and run the entire length of the tie-down area (anchor cables), some operators allow pilots to use the hook method—you pull the tie-down chain tight, place the hook through the chain link that's closest to the wing's tie-down ring, and then attach the hook to the tie-down ring. If you do this properly, neither the chain nor the wing's tie-down ring will rest on the hook's locking pin. The top of the hook will rest on the tie-down ring; the bottom of the hook will hold the chain. The locking pin is free.

FLIGHT OPERATIONS

GENERAL

Safety and aviation are inseparable.

- Do not save the concept of safety for special occasions. The words *safety* and *aviation* are inseparable.

- Never operate an airplane contrary to the POH; never compromise SOPs or violate federal regulations unless your safety is in jeopardy.

- File a flight plan for every extended cross-country flight.

- After airplane maintenance has been performed, test fly the airplane before you carry passengers or fly at night.

Wear ear plugs or headsets when flying.

- Wear ear plugs or headsets when flying. The noise that you cannot hear is what causes permanent hearing damage.

- Always know your location in reference to a prominent geographical feature, airport, city, or navigation aid.

- Always monitor surface wind indications when taxiing and when flying. This advance knowledge can make the difference between a successful and unsuccessful emergency landing. The atmosphere is constantly changing, and the reported winds are historical information that seldom reflect the actual wind at your location.

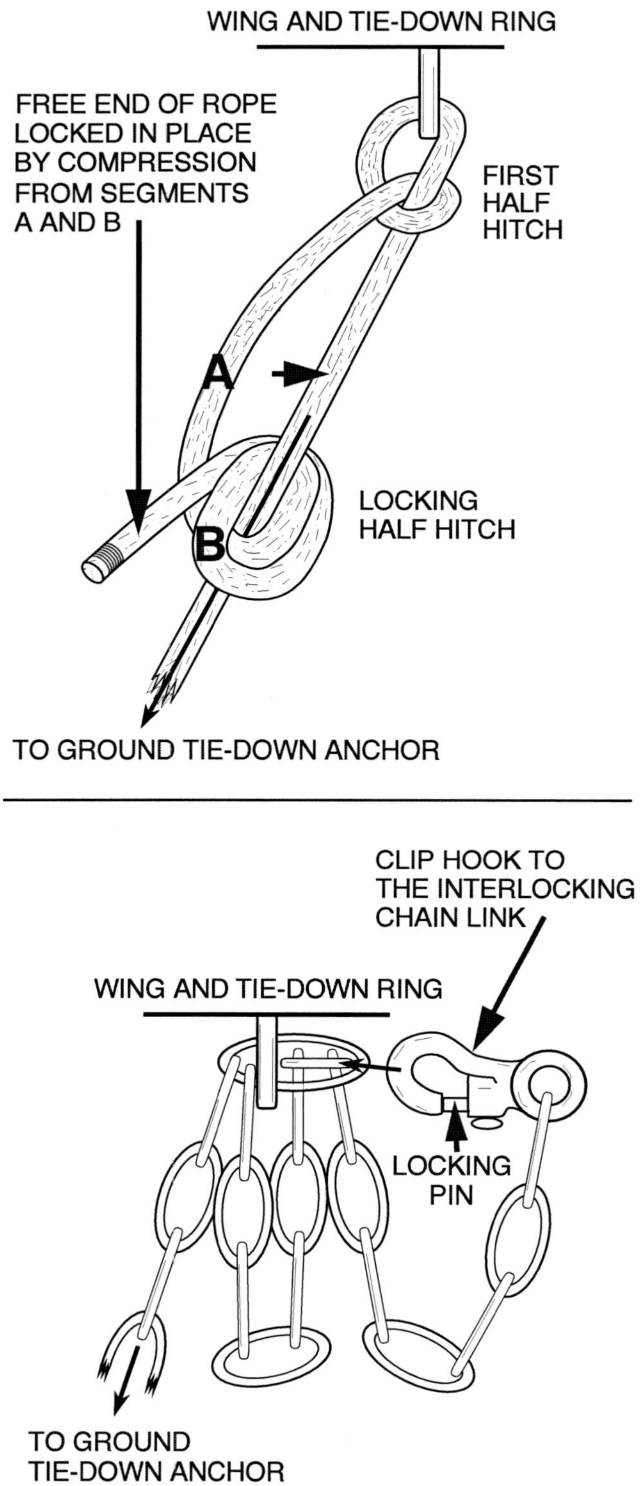

Rope Tie Downs

Use a locking half-hitch—a half-hitch with a second loop. Tie a regular half-hitch close to the tie-down ring; then tie the locking half-hitch within a foot of the first half-hitch. After forming the second loop, give the rope's free end a hard pull so that it wedges and locks itself between rope segments A and B. If a stiff, synthetic rope is used, a third half-hitch will be required below the locking half-hitch to hold the rope in place.

Chain Tie Downs

Use interlocking links. Remove the chain's slack and push one link through the wing's tie-down ring. Place a link from the chain's free end through the portion of the first link that protrudes beyond the tie-down ring. Attach the hook to this link. The chain now takes all the stress, not the hook, and when the hook shifts its position, the stressed chain will never pull against the hook's locking pin.

Some tail tie-down points—such as those found on Piper aircraft— have a large opening. If using rope, use the locking half-hitch method. If using chain, pull *two* strands of chain through the opening, and insert the chain link that is common to both strands through a link in the chain that is attached to the ground. Now, use the hook to secure that link. Again, all stress is on the chain, not the hook.

Rope and Chain Tie-Downs Attached to Airplane

- Always have a suitable forced landing area in mind.

Always clear your airspace for traffic.

- Avoid midair collisions. Clear your airspace. Know how to look, where to look, and when to look.

How To Look Use proper eye movement. During the day move your eyes from one 30-degree sector of airspace to the next, pausing for 5 to 10 seconds at each sector. At night move your eyes slowly across all sectors without stopping.

Use relative motion to evaluate collision potential. A collision will not occur if an aircraft is moving with respect to a fixed point on your airplane. A collision will occur if that aircraft seems to appear motionless.

Clean windows are important for collision avoidance. If two 180-MPH airplanes are converging head on, each would appear as a fixed spot on the respective windshields. If recognition is made at three miles, the pilots have only 30 seconds to react. Government studies have shown that 12.5 seconds are required—1.1 to see and recognize the object, 5.0 to evaluate collision course, 4.0 to decide on course of action, and 2.4 to start turning.

Where To Look Where the threat is greatest—the airspace within three miles of your airplane. I call the airspace that is ahead of the airplane the *cone of terror.*

When scanning during daylight hours, focus on a distant point every 30 seconds. This is done to overcome a physical limitation: When you stare into space, your eyes eventually return to their normal resting level and focus only a few feet ahead.

75 percent of all midair accidents occur within five miles of an airport and within 3,000 feet of the surface.

Pay particular attention during prolonged climbs and descents, prior to turning, and when flying near airports and radio navigation stations. In high wing airplanes, always raise the inside wing to clear the airspace before turning. The FAA once said that 75 percent of all midair accidents occur within five miles of an airport and within 3,000 feet of the surface.

Students must call out "clear right" or "clear left" prior to turning.

Clear each turn. Student pilots must call out "clear right" or "clear left" prior to turning so that the instructor knows that this important habit pattern has been established.

Clear the practice area. Prior to flying a maneuver or series of maneuvers, make at least two 90-degree, steep-banked turns to clear the airspace around, above, and below you.

When your airplane's shadow can be seen on the ground, it can provide an important reference. If another aircraft's shadow is observed in close proximity, stop what you are doing and look for that traffic.

Eighty percent of your time should be spent looking outside the cockpit.

When To Look Always! You must develop a good scan. A NASA study showed that competent pilots continually repeat this scan sequence: four seconds inside the cockpit; sixteen seconds outside the cockpit. In other words, eighty percent of your time should be spent looking outside the cockpit.

- Continually divide your attention. Never stare at one element. This is a hard habit to develop, but it is a critical piloting skill.

Here is a typical sequence: Outside for traffic and attitude. Inside briefly. Outside for traffic and ground reference points. Inside briefly. Outside for traffic and forced landing sites. Inside briefly. Outside for traffic and surface wind indications. Inside briefly, and then continue the sequence.

- Know who is flying the airplane. Use verbal commands when control transfer occurs. The pilot taking control states, "I have it." The pilot relinquishing control states, "You have it."

- Listen to the airplane; feel the airplane. Relying only on vision is not enough. Good pilots use all of their senses.

- When landing, remain on the runway centerline until the airplane decelerates to taxi speed.

- After landing, do not stop until you are clear of the runway's hold-short lines.

- Never clean up the airplane—move secondary controls or switches—until you have reached taxi speed and are clear of the runway. At busy airports, team work is required between all pilots and ATC in order to keep traffic moving. Call for taxi instructions before you clean up the airplane.

HEADWORK

Be professional, stay informed.

• Be professional, stay informed. Corporate flight departments, military units, and airlines keep their pilots updated at all times. General aviation pilots must do this on their own, and I cannot overemphasize the importance of joining the Aircraft Owner's and Pilot's Association—www.aopa.org.

Membership gives you access to an incredible amount of safety, operational, and training information on their extensive website as well as a subscription to the *AOPA Pilot* magazine. Student pilots receive a free six-month subscription to the AOPA's *Flight Training* magazine.

Always have an out!

• Always have an out! Flying is risk management, and aviation weather reports are nothing more than snapshots of a constantly changing environment. You must have a no-risk plan of action available just in case your intended plan becomes questionable or unacceptable. In other words, Keep your options open. Never commit to a single course of action that eliminates all viable options.

• Three things are constantly running through a good pilot's mind when flying: What's my location? Where's the wind? Where will I land if the engine quits?

• "Judgment comes from experience, and experience comes from poor judgment." [author unknown]

• You may be the world's greatest optimist when both feet are on the ground, but in the air, you must be a pessimist.

Be a pessimist. Question everything.

Would you accept the following statements? "I topped off your tanks with 100 octane, low-lead avgas." "Forecasts indicated nothing less than 3,000 foot ceilings and 4 miles visibility throughout the period." "You won't need the adjoining chart because your destination airport is on the edge of this chart." "The GPS receiver's data base is still good." No, you would never accept such statements at face value.

People are well intentioned, but with respect to aircraft safety, you the pilot are the last line of defense. Validate all information that you receive.

• Never make assumptions. Base your decisions on solid knowledge.

Do not let Murphy surprise you.

• Expect the unexpected. Anticipate any situation that could affect planned actions. Ignoring the unexpected leaves you unprepared to cope with the realities of flight. Do not let Murphy surprise you. This is in reference to Murphy's Law which states, "Anything that can go wrong will go wrong." Examples:

An extended meeting delayed your planned departure, so you must fly home at night, something you rarely do. The coastal destination was forecast to be VFR, but very few lights are visible as you approach the area. You suddenly notice a fuzzy glow surrounding the wingtip navigation lights. You have entered a cloud. Fortunately, you received enough instrument training to execute a carefully managed 180-degree turn that returns you to VFR conditions. How widespread is this unforecast weather? What airports are still VFR? How much fuel is left? Exactly where are you?

You are approaching your destination airport with minimum fuel due to stronger than forecast headwinds. Upon calling the tower, you are advised that the airport is closed due to an airplane that landed gear-up. The nearest airport is at least thirty minutes away.

• To fly safely, you must be a pessimist and develop your own rules for coping with the unexpected. Examples:

Always plan on an engine problem after takeoff.

Never assume that the weather will continue as forecast.

Never assume that your planned navaids will be working just because the FSS briefer said, "No pertinent notams."

Stick to your plan of action.

• Stick to your plan of action. Your destination airport is fogged in but it is forecast to clear up at your estimated time of arrival. You planned an alternate airport just in case the forecast was wrong. Your destination is still fogged in when you arrive, but when you call flight service, you are told that the visibility should start improving at any time. Ignore that opinion! Go to your alternate and refuel.

- Always note the time of takeoff.

Never rely on the fuel gauges. Use time aloft and known fuel consumption.

- Always base fuel consumption on time aloft and known fuel consumption, not the fuel gauges.

- Know the *actual* endurance limit for the airplane and the engine leaning procedure for your planned fuel consumption.

- Every fifteen minutes, check the heading indicator's alignment and your fuel status. It is normal for a heading indicator to drift fifteen degrees per hour, and that does not include the instruments normal precession errors.

- Always know where you were, where you are, and where you are going.

- Always look at a control or switch before moving it. Know what it does, how it does it, and why you want to move it. When you do move it, confirm that the desired action took place.

- Do not fixate. Keep dividing your attention between flying the airplane, navigating, communicating, and watching for traffic.

- Eliminate distractions and maintain an alert mental state.

Do not let self-confidence turn into complacency.

- Maintain confidence through practice, self-study, and recurrent training. Do not let self-confidence turn into complacency—a proven pilot killer.

- The common trait of all survivors is common sense.

- To avoid making all the mistakes yourself, learn from others. Study accident reports, hangar fly with other pilots, read aviation magazines, and attend safety seminars.

- "Flying is like playing cards. You always play with the same 52 cards, but you are never dealt the same hand twice." [Captain Len Morgan, Braniff Airlines, retired, *Flying* magazine, December 1990.]

REVIEW TEST (answers—after page Appendix - 7)

1. Define the term "standard operating procedures."
2. The final authority for aircraft operation is?
3. If required during preflight or postflight, how do you move the propeller?
4. What hazard exists if you pump the throttle to prime a carbureted engine?
5. How can spark plug fouling be minimized while taxiing?
6. If you do not have an EGT gauge, how do you lean the engine during cruise flight?
7. What indicates the formation of carburetor ice?
8. When carburetor ice starts to occur, how much carburetor heat should you use?
9. What ground check determines whether or not you can use idle power in flight?
10. Prior to engine shutdown, what two checks must be made?
11. What pilot self-checking procedures are used to prevent errors?
12. What pilot self-checking procedure is mandatory?
13. What rule must be followed when initiating radio communications?
14. Can you acknowledge an ATC instruction by double-clicking the push-to-talk button?
15. What must occur before an aircraft is refueled?
16. When cleaning plexiglass windows, what hand motion should be used?
17. What factors can cause fuel capacity to vary?
18. When walking out to preflight the airplane, what should you observe?
19. What is the most important preflight element?
20. When should strobe lights be operated?
21. When should landing lights be operated during daylight hours?
22. How should you manage fuel tank selection prior to takeoff?
23. If you inadvertently turn the magneto switch off during the magneto check, you must?
24. What must you check when you start to taxi? If only one brake worked, you would?
25. What commands are used when you transfer aircraft control to another pilot?
26. How do you control excessive taxi speed?
27. What instruments do you check while you taxiing to the run-up area?
28. When can you make powered "tail swings" in order to align the airplane for parking?
29. How should you tie-down an airplane?
30. What time increments do you use for scanning inside and outside the cockpit?
31. What controls can you activate while on the runway after landing?
32. Flying is risk management. What is your primary operating rule?
33. How do you determine aircraft fuel status when airborne?
34. What error prevents pilots from properly dividing their attention?

CHAPTER 3

MANEUVERS FOR SOLO FLIGHT

FUNDAMENTAL KNOWLEDGE

Presolo instruction is the most critical training stage. It develops the basic insights that will forever impact your performance as a pilot. Success depends on your ability to study, to develop the proper mental and physical habit patterns, and to demonstrate proper awareness and judgment.

The time that it takes to solo is immaterial.

The time that it takes to solo is immaterial. Years ago, about ten hours of instruction was required, and that is still true if a young student learns in a simple airplane at a remote, uncontrolled airport. As your age or airplane complexity increase, or when training occurs at a busy, controlled airport, the time to solo can easily increase to twenty-five hours.

Soloing in minimum time in a simple airplane at a remote airport does not give that pilot an advantage over others. Pilots who learn to fly at busy airports get what they pay for. Their enhanced confidence, awareness, and skills—particularly with air traffic controllers—justifies the additional training time. To take full advantage of an airplane, every pilot must sooner or later attain that level of proficiency.

Excellent pilots know how to fly the wing by using specific attitudes and power settings.

Excellent pilots know how to fly the airplane's wing by using specific attitudes and power settings. To do this, you must memorize the wing and power settings for each performance objective—a direct application of the old axiom, "Attitude plus power equals performance." Wolfgang Langewiesche, the author of *Stick and Rudder*, called performance objectives the "gaits of flight," and I do likewise in this manual.

All gaits of flight and all maneuvers must be introduced and mastered *without* instrument reference. After that, you will be introduced to instrument reference in order to fine-tune your performance. Get another instructor if your instructor cannot or will not teach flying in this manner.

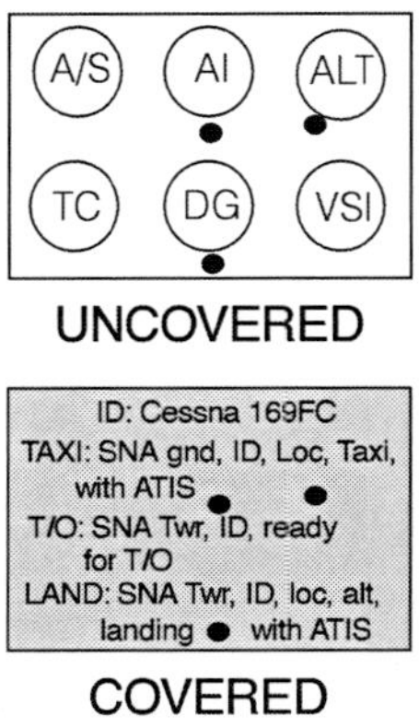

File-Folder Instrument Cover

To start, the flight instruments are covered as illustrated on the left. I place half of a manila file folder over the instruments and mark the location of the heading indicator, attitude indicator, and altimeter knobs. After I cover those marks with masking tape for reinforcement, I use a pencil or pen to punch holes at each location so that I can secure the cover to the instrument panel at those points. For student reference, I write the airplane's tail number and three basic communication sequences—taxi for takeoff, takeoff, and landing—on the cover with a felt tip pen.

THE FIVE SENSES

You cannot achieve the proper level of awareness when flying unless you fly relaxed and use your five physical senses properly—vision, hearing, touch, kinesthesia, and smell.

Vision for airplane attitude and relative motion

Fixating your sight on only one reference is a serious flying error.

You evaluate attitude by observing the airplane's wing and nose positions; you evaluate relative motion by observing the movement of the windshield's framework in relation to the Earth's surface. Are pitch, roll, and yaw changing at the desired rate or remaining constant? Is the airplane drifting laterally, climbing, or descending? There are always two or more references that you must scan for correct performance. To scan means that you continually divide your attention between such references. Fixating on only one reference is a serious flying error.

I cannot overstress the importance of learning to use your vision properly. To do this, you *must* change deeply ingrained, earthbound habit patterns.

When I ask students or licensed pilots how to fly a given maneuver, I usually get a correct answer. In the air, however, that maneuver is often flown incorrectly. Why does that occur? Because their eyes were not in the right place. They did not see the onset of an unacceptable deviation—and that's the primary reason why I wrote this manual.

Hearing for airspeed and engine speed

Listen to the airplane.

Airflow noise is an indirect indication of aircraft speed. Engine noise is an indirect indication of engine speed. Both tell you what the airplane and the engine are doing.

Touch for control pressures and attitude response

Feel the control pressures, and evaluate attitude response.

Control pressures and the resulting attitude response are an indirect indication of aircraft speed. High speed generates firm control pressure and rapid attitude response to control inputs; low speed generates light control pressures and slow attitude response to control inputs.

Kinesthesia for movement

The sensation of movement (kinesthesia) provides an indication of acceleration or deceleration with respect to the linear and rotational motion that pilots must be aware of during all phases of visual flight. You must, however, ignore these sensations during instrument flight.

Smell for odors

Smell the airplane.

Your nose sounds the alarm for fuel leaks, overheated electrical components, and exhaust leaks. Don't ignore the initial warning because your sense of smell tires—after a few minutes, you cannot detect previous odors that still exist.

AIRPLANE ATTITUDE

Attitude refers to the airplane's orientation in relation to the Earth's surface, and it is referenced in terms of pitch and bank. You refer to pitch attitude as nose high, level, or nose low; you refer to bank attitude as left bank, wings level, or right bank.

Attitude and power are inseparable.

Proper attitude and power awareness are the secret to flying with the proper level of self-confidence. Your first airplane management responsibility is to continually monitor attitude and power—they are inseparable.

While you monitor attitude and power, you must evaluate six types of motion that occur when flying. The first three are rotational: pitch, roll, and yaw. The last three are linear: forward, lateral (wind drift), and vertical (updrafts and downdrafts—also called risers and sinkers).

Warning! All your life you have looked straight ahead while monitoring forward linear motion. When you fly an airplane in a nose-high attitude—liftoff during a takeoff, climbs, stalls, slow flight, and the landing flare—you must look to the side of the nose cowl so that you can see enough of the Earth's surface to properly evaluate the six types of motion.

Attitude	Performance Objective	Power Setting
P 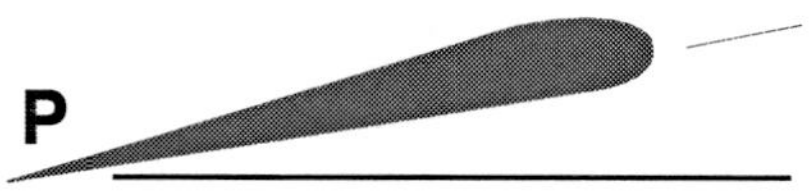	Best Angle Climb	Full Power
I 	Best Rate Climb	Full Power
T 	Cruise Climb Slow Cruise	Climb 2,000-2,200 rpm
C 	Accelerate Cruise Best Glide Slow Cruise Descent Power-On Landing Approach	Takeoff Cruise Idle 1,400-1,500 rpm 1,300-1,400 rpm
H 	Cruise Descent Power-Off Landing Approach	Cruise Idle

Pitch Performance

PITCH PERFORMANCE

Five pitch attitudes are used for normal flight in modern, normal category airplanes that have wing flaps.

Pitch has five letters that represent the five pitch attitudes required for normal flight. Three of these attitudes are nose high, one is nose level, and one is nose low. I call them the P, I, T, C, and H attitudes, and they represent specific increments of pitch change as illustrated on the opposite page.

Each performance objective requires a specific attitude and power setting. If you memorize these relationships, you will know the secret for flying modern, light airplanes because all of their wings work in the same manner.

To observe these attitudes, take any airplane and compare the imaginary chordline that is near the wingtip to the horizon while you perform each performance objective. The chord line is the imaginary line that connects the wing's trailing and leading edges. It makes no difference where the wing is attached to the fuselage—high, mid, or low.

To establish pitch attitude, you use the chordline and the horizon. To maintain pitch attitude, you use either the chordline or the airplane's nose and the horizon. If the nose moves up or down in relation to the horizon, pitch attitude is changing.

Unfortunately, some instructors teach students to use only the nose for establishing the pitch attitude. Students are told to place a specific nose-cowl reference on the horizon. When these students transition to another type of airplane, they are lost because that reference is missing.

To be an excellent pilot, you must learn to fly the wing!

When you learn to establish pitch with the wing's chordline, you will discover that doing the same thing in other airplanes results in total success. The wing is what makes flight possible. To be an excellent pilot, you must learn to fly the wing!

Does this technique yield perfect results? Absolutely not, but you will be close to your desired objective and you will quickly achieve it without compromising your safety. After you master flying the wing using attitude, power, and visual reference points, your instructor will introduce instrument reference. Now you can obtain exact results *without staring* at the flight instruments—another extremely serious pilot error.

The flight instruments are not your first priority, they are your third priority.

Instrument reference means that you glance at the appropriate instrument and then look outside the cockpit to make your correction. A government study stated that competent pilots spend sixteen seconds looking outside the cockpit for every four seconds they spend looking inside the cockpit. Always remember this fact: The flight instruments are not your first priority, they are your third priority.

Attitude	Performance Objective	Power Setting
B 	Level Flight	Any
A 	Shallow Bank Turn	Any
N 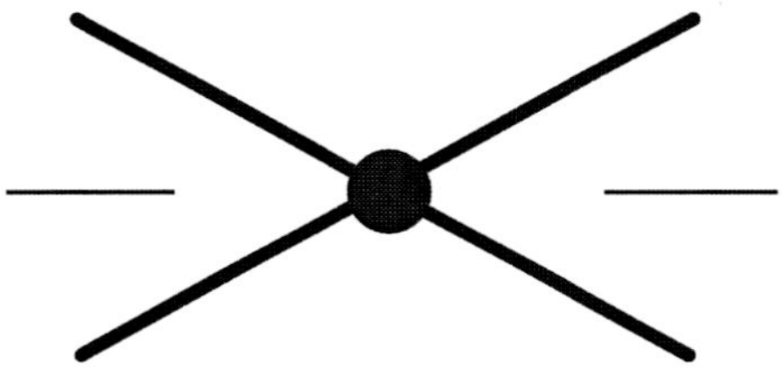	Medium Bank Turn	Any
K 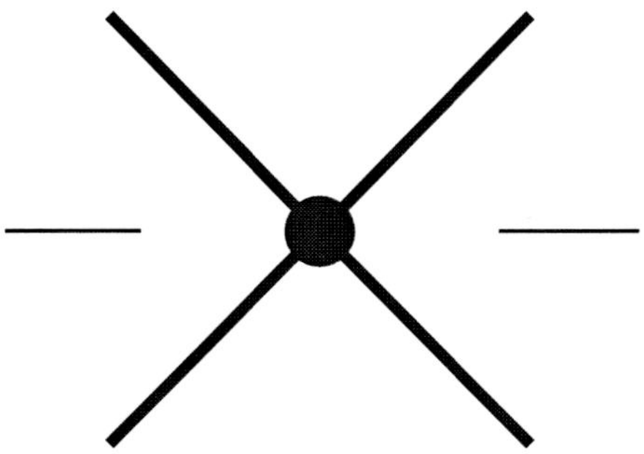	Steep Bank Turn	Any

Bank Performance

Your first two priorities are attitude and power.

Your first two priorities are attitude and power, and while doing that, you must divide your attention in order to evaluate airplane motion, watch for traffic, and monitor radio communications. When that level of proficiency occurs, your self-confidence will be at the proper level.

BANK PERFORMANCE

Four bank attitudes are used for normal flight.

Bank has four letters that represent the four bank attitudes required for normal flight. The illustration on the opposite page shows those attitudes: three are banked and one is wings level. I call them the B, A, N, and K attitudes, and they represent specific increments of bank change. To observe them, take any airplane and compare the nose position to the horizon during wings-level flight, a 15-degree bank, a 30-degree bank, and a 45-degree bank.

To establish bank, you use the airplane's nose and the horizon. To maintain bank, you use the nose in relation to the horizon or use the lower wingtip's motion in relation to the Earth's surface. That wingtip will rise or drop if bank angle changes.

	Imaginary Chordline	Nose
PITCH	Use to establish and maintain pitch attitude	Use to maintain pitch attitude
BANK	Use to maintain bank attitude	Use to establish and maintain bank attitude

Establishing and Maintaining Pitch and Bank Attitudes

NEUTRALIZING THE CONTROLS

Neutralize the controls does not always mean the removal of a flight control input.

When instructors use the word *neutralize*, it refers to flight control action that is required in order to maintain attitude. In this section you will learn that neutralize the controls does not always mean the removal of a flight control input.

Ailerons

When you rotate the yoke from its neutral position, the ailerons are displaced from their neutral position—streamlined with the wing—and the airplane will roll indefinitely. When you return the yoke to its neutral position, the airplane will stop rolling, but there are exceptions.

During shallow- and steep-bank turns, bank attitude will change if the ailerons are neutralized.

During shallow-bank turns, the airplane will return to a wings-level attitude when the ailerons are neutralized because of aerodynamic stability that all airplanes possess. During steep-bank turns, however, the bank angle will increase, because the outside wing is moving faster than the inside wing and is therefore generating more lift. During slow speed flight with high power settings and neutralized ailerons, the airplane will roll to the left due to engine torque and other factors. In all three cases, the ailerons must be displaced slightly in order to maintain the desired bank angle.

Rudder

When you displace the rudder from its neutral position—streamlined with the vertical stabilizer—the airplane yaws until the rudder force is balanced by opposing aerodynamic forces. To neutralized the rudder, you return it to a position that maintains the proper yaw rate but, contrary to the ailerons, this is rarely the neutral position.

When flying straight—heading is constant—the airplane should not yaw, so right rudder is required for climbs and left rudder for descents. When turning, the yaw rate must be proportional to bank angle and airspeed. After you establish a turn, the rudder cannot be returned to the neutral position, because it must remain aligned with the curved flight path.

After you establish the desired attitude, use rudder to eliminate adverse yaw.

Adverse yaw results from normal aileron displacement and from aileron displacement that is used to counteract the following: roll from engine torque, unequal thrust on the propeller disc (P-factor) when the airplane is in a nose-high or nose-low attitude, the corkscrew effect (spiraling propwash hitting the left side of the vertical stabilizer), and gyroscopic precession of the propeller. These forces vary with different combinations of power and airspeed. Have no fear, the solution is simple—use elevator and ailerons to maintain the desired attitude; use rudder to eliminate adverse yaw.

Elevator

When you displace the elevator from its neutral position—streamlined with the horizontal stabilizer—the airplane will pitch up or down. It stops pitching when the elevator force is balanced by opposing aerodynamic forces. To neutralize the elevator, you return it to a position that maintains the proper pitch attitude, and like the rudder, this is rarely the neutral position.

TWO-STEP CORRECTIONS

Purpose

Fixation is an extremely serious pilot error.

To eliminate fixation, an extremely serious pilot error that prevents scanning. You must continually divide your attention and cross-check the applicable items from the time you start the engine until you shut it down.

Two-Step Control-Input Rule

When an attitude or power correction is required, make one that is obviously excessive. When it starts to work, remove half of that input.

- If airspeed is incorrect and a pitch correction is required, make an excessive pitch change. When it starts to work, remove half of that correction. If airspeed is incorrect and a power correction is required—a common occurrence on final approach—make an excessive power change. When it starts to work, remove half of that correction.

When you use engine power in this manner, do not look at the tachometer. Keep dividing your attention and index the throttle. You index power corrections by using either engine sound or hand movement.

Indexing the throttle does not imply jerky power changes.

With engine sound, note the sound before and after you make the correction. When the correction starts to work, set the power so that the sound is at an intermediate level.

With hand movement and a panel-mounted throttle, place your extended index finger against the panel and hold the throttle knob between the fourth and fifth fingers and the palm of your hand. With a quadrant-mounted throttle, rest the outside heal of your hand on the quadrant.

- If lateral drift occurs due to a crosswind during landing, establish a bank angle that will obviously stop the drift. When it starts to work, remove half of that correction.

The two-step, control-input rule is a mandatory procedure. A serious error occurs when a pilot, realizing that a correction must be made, adds a little, adds a little more, and then adds even more. That pilot is an *adder* and is therefore snake bitten, because an Adder is a venomous snake. Adders fixate on the problem and stop scanning. Never let that happen!

Two-Step Correction Rule

Stop the deviation from increasing. That's step one. Then eliminate it.

When the airplane is deviating from a specific objective, do not attempt to put it back where it belongs with one correction. First, stop the deviation from increasing, and then establish a new attitude or power setting that will eliminate the deviation. If you attempt to do this with just one correction, you will overcontrol the airplane and fixate on that problem.

Altitude control during cruise is an example. When the airplane deviates from the desired altitude, stop the deviation by establishing a pitch attitude that maintains the existing altitude. Now, make a slight pitch adjustment that will slowly return the airplane to the desired altitude. If a significant attitude change is required, you must also change power in order to maintain airspeed.

As you gain experience, two-step corrections become smaller, but they never go away.

As you gain experience, two-step corrections become smaller and smaller, but they never go away. If someone called me the world's best pilot, I would say, "That's impossible, I'm always making small corrections."

TRIMMING

Purpose

To relieve control pressures during stabilized flight in order to fly hands-off whenever necessary. Trim requirements change when airspeed changes due to attitude or power changes.

Execution

Larger airplanes have trim controls for all three primary control surfaces—the elevator, rudder, and ailerons. Basic training airplanes, however, have only an elevator trim control in most cases, and when rudder or aileron trim is required, continual rudder or aileron control pressure must be applied.

The trimming sequence is elevator, rudder, and ailerons.

You must follow this sequence for trimming an airplane: elevator, rudder, and ailerons. This sequence applies even though the airplane may not have rudder and aileron trim. If the airplane has those controls, they should be neutralized before you start the trimming sequence, and the fuel quantity in each wing tank should be equal.

To trim the elevator, you hold the pitch attitude constant with the elevator and remove the elevator pressure with the trim control. When you think the elevator is in trim, let go of the yoke and watch the airplane's pitch attitude using the nose reference. If pitch changes, reestablish the desired attitude with the elevator and apply additional elevator trim.

If learning to fly in an airplane with electric trim, do not use it.

If you are learning to fly in an airplane that has electric elevator trim, do not use it. Trim the airplane manually. To use electric trim, you must grip the yoke firmly in order to activate the trim switch, which makes you lose your sense of control feel and makes learning more difficult.

Once you have mastered manual trimming, you can use the electric trim. To do this, you *beep* the trim switch once or twice, and then release the yoke and watch for an attitude change. Repeat the process until attitude remains constant. Beeping means that you make extremely short inputs by snapping your thumb across the trim switch. Prolonged electric trim inputs cause overcontrolling, and this is not recommended unless the elevator control force is extremely high.

To trim the rudder, you must keep your hand on the yoke in order to keep the wings level during this entire step. Stop yaw with rudder input, and then remove the rudder pressure with rudder trim. Release the rudder pedals to check for yaw. If it occurs, stop it with rudder input and apply more trim.

To trim the ailerons, you use aileron trim to remove the aileron force that was required to hold the wings level while you trimmed the rudder. Release the yoke and check for roll. If it occurs, stop it with aileron input and apply more trim.

To stay in trim when the airplane does not have rudder and aileron trim, you must apply continuous control inputs. But, if the airplane has fixed trim tabs on the rudder and on one aileron, you can bend them when you are on the ground so that the airplane stays in trim during cruise flight, the most common flight mode. This process may take several flights.

Attempting to fly the airplane with the trim controls is a serious error.

If you attempt to fly the airplane with the elevator trim control, you will get *behind* the airplane. In other words, you will lose your control-input awareness and increase your workload. We fly airplanes with the primary controls and use the trim controls only to eliminate control pressures.

Do not rush the trimming process. Nothing is instantaneous when it comes to airplanes, and you must give the airplane time to settle down and respond to your inputs.

If you have difficulty trimming an airplane, it may be due to the control system's internal friction. Assume that you have trimmed the elevator as best you can, but the nose still wants to rise. Using very light pressure, move the yoke slightly forward. The control system friction will probably keep the yoke in that position, and the nose will stop rising.

Your instructor will use the terms *fine trim* and *rough trim*. You fine trim the airplane when you follow the previous procedures. You rough trim when you start a maneuver and the control force becomes excessive. The airplane is not in stabilized flight, and you are not trimming for hands-off flight. You are trimming only to remove the excessive control force.

When busy, you can temporarily control the airplane without touching the yoke.

Airplanes can fly themselves with minimum pilot input, and you must learn to do this when your workload is high—tuning radios, programming a GPS receiver, locating and reading navigation charts, and performing the lost and the diversion navigation procedures. To do this, take your hand off the yoke, use rudder to maintain bank angle if turning or heading if flying straight, and use power to maintain altitude. If bank is incorrect, step on the high wing's rudder pedal to regain the desired bank angle—left wing high, left rudder; right wing high, right rudder. If the airplane starts climbing or descending, increase or decrease power slightly in order to stop the deviation—this is called pitch trimming with power.

When you fly hands-off and maintain your heading, a slight bank may occur and the inclinometer (the ball at the bottom of the turn coordinator) may be slightly off center. This is acceptable because you must maintain heading. When your workload returns to normal, you will put your hand on the yoke and fly the airplane in the normal, coordinated manner.

When an airplane has rudder trim but no aileron trim, a common error can occur when you fly hands-off after you have trimmed the rudder. If some rudder pressure is still required to maintain bank angle due to the lack of aileron trim, do not change the rudder trim. That action will untrim the rudder and hands-off flight will become difficult.

The step-on-the-high-wing rule for hands-off flight is an extension of the step-on-the-ball and the step-on-the-runway-centerline rules, and you should use it when flying in turbulence. If you step gently on the rising wing's rudder pedal, yaw will be minimized, but if you use ailerons to level the wings, yaw will continue. That means airsickness bags for your passengers. Yuck!

Never say that you fly the airplane with just rudder and throttle. You can say, however, that you temporarily control the airplane with rudder and throttle.

Caution! Never say that you fly the airplane with just rudder and throttle. Airplanes are flown with rudder, aileron, elevator, and throttle in a coordinated manner. However, when the airplane is stabilized and you need both hands for a specific task, you can say that you temporarily control the airplane with rudder and throttle.

SEAT ADJUSTMENT

- Do not overlook this important step. Adjust the seat horizontally so that you can easily apply full rudder while your heels remain on the cockpit floor.

- Adjust the seat-back tilt so that you are comfortable when you hold the yoke in it's neutral-elevator position.

- Adjust the seat vertically until you just barely see the top of the nose cowl when you look across the top of the instrument panel's glare shield. That position is not too low, as you will learn when you start using the proper sight pictures.

THE FIRST THREE LESSONS

No instruments, attitude plus power, and the proper visual references. It's that simple.

The first three flight lessons of the Skyroamers® Private Pilot Flight Training Syllabus teach three important objectives: You do not need the flight instruments to maintain proper aircraft control. Attitude plus power equals performance. And, by using the proper visual references, you can immediately detect and correct unwanted deviations.

During these short-duration lessons, your instructor will cover the flight instruments, start a meaningful instructor-student relationship, and eliminate any apprehension that you may have about airplanes or flying. You will learn faster and fly better if you are relaxed during your lessons.

The first lesson is a demonstration flight. It introduces you to the airplane's controls, familiarizes you with the local area, and dispels myths that you may have regarding airplane operation. Your instructor will demonstrate proper control usage, aircraft stability when the controls are released, glide capability when the engine is idled, and the simplicity of the stall and the stall recovery by performing a flaps-up, power-off stall with a power-off recovery.

The second lesson reviews all factors that affect pitch: elevator input, airspeed, power, bank angle, flap extension and retraction, and the elevator trim control. Basic pitch performance objectives are introduced: best-rate climb, cruise climb, cruise, slow cruise, and cruise descent.

The third lesson reviews all factors that affect bank and yaw: aileron and rudder inputs, airspeed, power, bank angle, and the rudder trim control if installed. Basic bank performance objectives are introduced: zero bank for straight and level flight, and shallow, medium, and steep bank turns.

Flying is simplified when you concentrate on attitude and power.

By the time you complete lesson three, the complexity of the pitch, bank, and power relationships may decrease your enthusiasm. Fortunately, your instructor will rescue you by saying, "Yes, there are many complex factors that influence airplane control. The complexity of these relationships, however, can be eliminated by following one simple rule: Concentrate on using the controls to maintain the required attitude.

The rudder controls yaw, the elevator controls pitch, and the ailerons control bank angle. But, while you are establishing the desired attitude, you must simultaneously set an appropriate power setting. That's how you quickly achieve the desired performance objective without compromising safety."

For effective instruction, an excellent instructor-student relationship must exit.

During these lessons, your instructor will try various techniques in order to establish the best instructor-student relationship. The instructor will ask questions, remain silent, talk softly, talk loudly, get the you to talk, be demanding, be easy, and evaluate your responses. By doing this, the training process becomes more effective. Be sure to tell your instructor about your background and what type of instruction you enjoyed the most during previous learning endeavors.

The fourth and subsequent lessons cover the maneuvers required for solo flight, which are presented next in their general order of introduction. The common errors for each maneuver are discussed the FAA's *Airplane Flying Handbook*, an excellent reference that you should purchase. I wrote this manual to supplement that manual because of a significant pilot error that it does not emphasize—what you must do with your eyes in order to be a safe, competent pilot.

Use carburetor heat before you set idle power if your airplane has a carburetor.

For proper flap retraction procedures, refer to your POH.

The maneuvers in this manual are discussed in terms of purpose and execution—how to fly the maneuver with respect to attitude, power, trim, and vision. CFI notes are included in order to help flight instructors teach properly. When I tell you to set idle power, apply carburetor heat if your airplane has a carburetor and icing conditions exist. When I tell you to select approach flaps for acceleration, refer to your airplane's POH. Some POHs specify that you initially retract only the last increment of flap extension.

Caution! Flight instructors can teach you to fly. They cannot, however, teach awareness, good judgment, multitasking, common sense, or force you to study. If one or more of these elements cause you problems during flight training, your instructor will not let you solo and most certainly will not recommend you for a pilot certificate. Do not be offended. Your instructor is just ensuring your safety, complying with federal mandates, and avoiding liability exposure.

To become a flight instructor, a person must be a commercial pilot, receive extensive flight instructor training and a recommendation from an experienced instructor, and be thoroughly tested by the FAA. Once certified, the instructor must follow specific, regulatory mandates.

You, in turn, must have your instructor's endorsement in order to fly solo and be recommended for a pilot certificate or rating. The FAA—and the aviation industry in general—has no tolerance for incompetent instructors, and I have never known a flight instructor who would jeopardize his certificate by certifying a student who could not demonstrate proper knowledge, proficiency, awareness, judgment, and common sense.

BASIC AIRWORK

STRAIGHT AND LEVEL—CRUISE

Purpose

To maintain straight and level flight at cruise speed, hands-on and hands-off, for an indefinite period. To minimize the elapsed time between two points and develop a critical navigation skill—flying a constant heading and altitude.

Attitude

Establish and maintain the C-pitch attitude and the B-bank attitude.

Power

The value recommended in the pilot's operating handbook (POH), usually 75-percent power.

Trim

When airspeed and power stabilize, fine-trim the controls.

Vision

Two points must be scanned:

- Use the wingtip and the horizon to establish the pitch attitude and maintain bank attitude—when the wings are level, both wingtips will be an equal distance above or below the horizon.

• Use the nose and the horizon to establish the bank attitude, maintain pitch attitude, and detect yaw.

• To maintain heading, pick a reference point that is well ahead of you on your ground track. Keep that point and the airplane's nose in the same relative position—the angle between the airplane's heading and the ground track is the wind correction angle. As you proceed, continue to pick additional reference points that are well ahead of you on your ground track. If you continually refer to the heading indicator to maintain heading, heading control becomes difficult.

Evaluate the wingtip and nose attitude while you scan the surrounding airspace for traffic. Master this maneuver with your hands on and off the yoke. When you fly hands-off use the rudder to maintain bank angle if turning or heading if flying straight; use power to maintain altitude.

STRAIGHT AND LEVEL—SLOW CRUISE

Purpose

To maintain straight and level flight at reduced airspeed, hands-on and hands-off, for an indefinite period. Slow cruise airspeed is normally located just below the maximum speed for flap extension, V_{FE}—the top of the white arc on the airspeed indicator. You use it to enter a traffic pattern and fly the downwind leg and when cruising in the practice area.

Slow cruise—to cruise at reduced airspeed.

Attitude

Establish and maintain the T-pitch attitude and any bank attitude.

Power

Approximately 2,000 rpm.

Trim

When airspeed and power stabilize, fine-trim the controls.

Vision

Two points must be scanned.

• Use the wingtip and the horizon to establish the pitch attitude and maintain bank attitude.

- Use the nose and the horizon to establish the bank attitude, maintain pitch attitude, and detect yaw.

- To maintain heading, pick a reference point that is on your ground track, and keep that point and the airplane's nose in the same relative position. As you proceed, continue to pick additional reference points that are ahead of you on your ground track.

Evaluate the wingtip and nose attitude while you scan the surrounding airspace for traffic. Master this maneuver with your hands on and off the yoke. When you fly hands-off use the rudder to maintain bank angle if turning or heading if flying straight; use power to maintain altitude.

LEVEL TURNS

Purpose

To turn using shallow, medium, and steep bank angles while maintaining altitude. To stop a turn with the nose on a visual reference point, not a compass heading.

Attitude

Establish and maintain the C- or T-pitch attitudes (cruise or slow cruise) and the A-, N-, or K-bank attitude.

Power

Cruise or slow cruise.

Trim

Do not retrim the elevator during the turn. Yoke back pressure is tolerated because turns are short-duration maneuvers.

During a turn entry, drag increases as bank angle increases and this causes airspeed to decrease and elevator back pressure to increase. When you complete the turn, drag, airspeed, and control forces will return to their original values.

Vision

Clear the airspace for traffic before turning.

To execute a turn, you must look straight ahead and watch the airplane's nose and the horizon. With respect to pitch, keep the airplane's nose in the same position relative to the horizon during the turn. Do not look away from the nose reference until you have established the bank attitude and neutralized the ailerons.

Proper aileron and rudder coordination is indicated by maintaining a reasonable rate of turn for the existing bank angle and airspeed. Yes, you can refer to the inclinometer's ball, but you will become a much better pilot if you learn to evaluate turn rate by watching the relative movement between the airplane's nose and the Earth's surface. If the turn rate is incorrect for the existing bank angle and airspeed, correct it with the rudder. Turn rate increases as bank angle increases and as airspeed decreases; it decreases as bank angle decreases and as airspeed increases.

To stop a turn, you must again look straight ahead and watch the airplane's nose and the horizon. With respect to pitch, keep the airplane's nose in the same position relative to the horizon. Do not look away from the nose reference until you level the wings, neutralize the ailerons, and stop any yaw.

CFI Notes

When teaching turns, elevator and rudder coordination will be established faster if a small reference mark is placed on the windshield with a piece of tape, a self-adhesive paper dot, or a grease pencil. To do this, fly straight and level and have your student look straight ahead at the horizon, parallel to the airplane's longitudinal axis. Place the mark where the student's line of vision intersects the windshield. Now, have the student keep this mark on the horizon while turning.

CLEARING TURNS

Purpose

Clearing turns—to check for traffic before starting a maneuver.

To clear the surrounding airspace for traffic before starting a maneuver or series of maneuvers providing that series will allow continual observation of the surrounding airspace.

Execution

Use good judgment. Make at least two 90-degree, steep-bank turns. Make certain that your airspace is clear of traffic before you practice a maneuver.

Attitude
Establish and maintain the C- or T-pitch attitudes (normal or slow cruise) and the K-bank attitude.

Power
Cruise or slow cruise.

CRUISE CLIMB

Purpose
To gain altitude at a reasonable rate while maximizing forward visibility, engine cooling, and forward speed in order to reduce flight time. You establish a cruise climb when you reach a safe altitude after takeoff if you are flying to the practice area or on a cross-country flight.

Attitude
If airspeed is at or above the cruise-climb speed, establish and maintain the T-pitch attitude and any bank attitude. If airspeed is below the cruise-climb speed, maintain the C-pitch attitude, accelerate to the cruise-climb speed, and then establish the T-pitch attitude. At level-off, establish the C-pitch attitude for cruise or establish the T-pitch attitude for slow cruise if you are in the practice area.

Power
Climb power. If airspeed is above the cruise-climb speed, do not increase power until airspeed decreases to the climb speed. If airspeed is at or below the cruise-climb speed, increase power immediately. At level-off, let the airspeed accelerate to the cruise speed before you reduce power.

Trim
Rough trim the elevator as necessary. When airspeed and power stabilize, fine-trim the controls.

Vision
Three points must be scanned. Evaluate these references while you scan the surrounding airspace for traffic:

- The wingtip and the horizon are used to establish and maintain pitch attitude. With zero bank both wingtips will be an equal distance from the horizon.

• The side of the nose and the horizon are used to establish bank attitude and maintain pitch attitude.

• The Earth's surface and the horizon, framed by the side of the nose cowl and the windshield post, are used to supplement the first two references and evaluate roll, pitch, and yaw.

This last reference is critical. I call it the *pie-in-the-sky* sight picture, because you must look through the windshield's somewhat pie-shaped lower left or lower right corner. (Refer to the illustration on page 4 - 17.) You must use it whenever you are in a nose-high attitude, because if you attempt to look over the nose, visual references become almost non-existent and your performance will be unsatisfactory. I guarantee it!

Remember this: The attitude indicator tells you what the airplane is doing, the other five flight instruments tell you what the airplane has done, and the outside-the-cockpit visual references tell you what the airplane is going to do. If you know that, you know the secret for maintaining positive airplane control and developing the proper level of self-confidence.

BEST-RATE CLIMB (V_Y)

Best-rate climb (V_Y)—to gain altitude as fast as possible.

Purpose

To gain altitude as fast as possible—in minimum time. To quickly reach the safe altitude after takeoff. The safe altitude is the altitude that gives you a forced landing option in the event of an engine failure.

Attitude

Establish and maintain the I-pitch attitude and the B-bank attitude. If present airspeed is at or above the best-rate climb airspeed, V_Y, pitch attitude selection is immediate. If present airspeed is below V_Y, maintain the C- or T-pitch attitude until the airplane has accelerated to V_Y. To terminate the climb, establish and maintain the desired cruise-pitch attitude.

Power

Full power. If airspeed is currently above V_Y, do not increase power until airspeed decreases to V_Y. If present airspeed is at or below V_Y, add power immediately. To terminate the V_Y climb, accelerate to the new speed before you reduce power.

Trim
Rough trim the elevator as necessary. When airspeed and power stabilize, fine-trim the controls.

Vision
Three points must be scanned. Evaluate these references while you scan the surrounding airspace for traffic:

- The wingtip and the horizon are used to establish and maintain pitch attitude. With zero bank both wingtips will be an equal distance from the horizon.

- The side of the nose and the horizon are used to establish bank attitude and maintain pitch attitude.

- The pie-in-the-sky sight picture is used to supplement the first two references and evaluate roll, pitch, and yaw.

CRUISE DESCENT

Cruise descent—to descend using the maximum forward speed.

Purpose
To maintain a comfortable rate of descent—500 feet per minute in an unpressurized airplane—and maximum forward speed. To minimize the flight time between two points while descending. This occurs when you return from the practice area or when you descend on a cross-country flight.

Attitude
Establish and maintain the H-pitch attitude and the B-bank attitude. At level-off, establish and maintain the cruise or slow-cruise attitude.

Power
Reduce as necessary to maintain cruise rpm—it will increase as airspeed increases in fixed-pitch propeller airplanes. Reduce power further if required by airspeed limitations or turbulence. At level-off, set cruise or slow-cruise power.

Trim
When airspeed and power stabilize, fine-trim the controls.

Vision
Three points must be scanned. Evaluate these references while you scan the surrounding airspace for traffic:

- The wingtip and the horizon are used to establish pitch attitude and maintain bank attitude.

- The nose and the horizon are used to establish bank attitude and maintain pitch attitude.

- The Earth's surface and the horizon, framed by the nose cowl and the windshield's framework, are used to supplement the first two references and evaluate roll, pitch, and yaw.

During level flight and descents you have a large view of the Earth's surface and the horizon, framed by the entire nose cowl and the windshield's framework. The pie-in-the-sky is used only when flying in a nose-high attitude—at liftoff during a takeoff, climbs, slow flight, stalls, and the landing flare.

SLOW-CRUISE DESCENT

Slow-cruise descent—to descend at slow-cruise airspeed.

Purpose

To lose altitude while maintaining slow-cruise airspeed. Used when maneuvering in the practice area.

Attitude

Establish and maintain the C-pitch attitude and the B-bank attitude. If airspeed is currently at or below slow-cruise descent speed, establish the C-pitch attitude immediately. If airspeed is currently above the slow-cruise descent speed, decelerate to this speed while maintaining altitude and then establish the C-pitch attitude. At level-off, establish and maintain the T-pitch attitude for slow cruise.

Power

Slow-cruise descent power. To determine this setting, start a descent using idle power. The engine is now being turned by the propeller which is being turned by the relative wind. Now, increase power until the you sense that the engine is turning the propeller. Listen to and remember the sound of the engine. This is the slow-cruise descent power setting, and it will be between 1,400, and 1,500 rpm. The rpm value will change between identical airplanes because tachometers are not that accurate at low rpms, but the engine sound will always be the same. At level-off, set slow-cruise power.

Trim

When power and airspeed stabilize, fine-trim the controls.

Vision

Three points must be scanned. Evaluate these references while you scan the surrounding airspace for traffic.

- The wingtip and the horizon are used to establish pitch attitude and maintain bank attitude.

- The nose and the horizon are used to establish bank attitude and maintain pitch attitude.

- The Earth's surface and the horizon, framed by the nose cowl and the windshield's framework, are used to supplement the first two references and evaluate roll, pitch, and yaw.

CLIMBING AND DESCENDING TURNS

Purpose

To change heading while climbing or descending. To begin and terminate climbs and descents while turning. To rollout on specific visual headings, not compass headings.

Attitude

Establish and maintain the appropriate attitude. Any combination of the five pitch attitudes and the three bank attitudes may be used. However, if you use the P- or I-pitch attitudes and the K-bank attitude, the airplane might stall.

When you bank an airplane, drag increases in direct proportion to the bank angle, and the increased drag will decrease the airspeed. Therefore, as you increase bank angle, you must decrease pitch attitude to prevent a loss of airspeed. This correction becomes critical when you make medium to steep-bank turns using the N- and K-bank attitudes.

Power

The power used during climbing or descending turns is identical to that used for straight climbs or descents. At level-off, set cruise or slow-cruise power.

Trim
Trim for the climb or descent, but do not retrim the elevator during turns. Back pressure is tolerated during turns.

Vision
Three points must be scanned. Evaluate these references while you scan the surrounding airspace for traffic.

- The wingtip and the horizon are used to establish pitch attitude and maintain bank attitude.

- The nose and the horizon are used to establish bank attitude and maintain pitch attitude.

- When climbing the pie-in-the-sky sight picture is used to supplement the first two references and evaluate roll, pitch, and yaw. When descending the Earth's surface and the horizon, framed by the nose cowl and the windshield's framework, are used to supplement the first two references and evaluate roll, pitch, and yaw.

To fly with confidence, you must use the proper visual references.

Are you annoyed by my continual repetition of each maneuver's visual references? Well, please realize that repetition is the key for establishing habit patterns, and you must quickly make these references habitual. Your old, ingrained, visual habit patterns that you developed while earthbound will conflict with those required for flying. Changing a person's earthbound habit patterns is a flight instructors most difficult task.

COORDINATION EXERCISES

ELEVATOR-THROTTLE COORDINATION

Purpose

When you can vary the power, pitch and power must be used together to maintain airspeed.

To perform two coordination exercises that require the proper use of pitch and power. During the first exercise you must maintain airspeed; during the second exercise you must maintain altitude. To realize that when power is fixed, pitch controls airspeed, but when you can vary the power—landing approaches and slow flight—pitch and power must be used together in order to control airspeed.

Execution for Constant Airspeed

Clear the surrounding airspace for traffic. Set climb power, maintain the best-rate climb speed, and use visual references to note the pitch attitude. Next, set idle power, maintain the best-rate climb speed, and again note the pitch attitude. Remember those two attitudes.

Continue to alternate between climb and idle power while you maintain the best-rate climb speed. Airspeed will remain constant only if you coordinate your inputs so that you reach your target power setting and target attitude simultaneously.

For constant airspeed, your hands move simultaneously in opposite directions.

Note that your hands move in opposite directions during this exercise. If your left hand moves the yoke aft, your right hand must move the throttle forward; if your left hand moves the yoke forward, your right hand must move the throttle aft. Or, if your right hand moves the throttle forward, your left hand must move the yoke aft; if your right hand moves the throttle back, your left hand must move the yoke forward.

Execution for Constant Altitude

Clear the surrounding airspace for traffic. Set climb power but maintain altitude. Approaching cruise speed, set idle power but, again, maintain altitude. Approaching stall speed, set climb power, maintain altitude, and repeat the exercise.

For constant altitude, your hands move in the same direction, but not simultaneously.

Note that your hands move in the same direction but not simultaneously. When power is increased, you must move the yoke forward slightly to compensate for the power-induced pitch-up, pause, and then continue to move the yoke forward as airspeed increases. When power is decreased, you must move the yoke back slightly to compensate for the power-induced pitch-down, pause, and then continue to move the yoke back as airspeed decreases.

CFI Notes

The previous exercises should be called elevator-throttle-rudder coordination because rudder must be used to maintain heading. This omission is deliberate to keep from diminishing the main objective of the exercises—elevator-throttle coordination, a subtle but critical skill that all students must understand and master. Poor rudder coordination during constant heading maneuvers is quite obvious and is easy to correct.

CONTROL-HARMONY EXERCISE

Purpose

To evaluate aileron, rudder, and elevator pressures in order to determine the airplane's control harmony (personality). To develop attitude awareness.

Control harmony refers to the control pressures that are required for coordinated flight. Some airplanes are light on the rudder and ailerons, but heavy on the elevator. Others are heavy on the elevator and rudder, but light on the ailerons. This exercise allows you to quickly determine these relationships, and it should always be performed during initial training and during aircraft checkouts.

A pilot's airplane requires equal pressure on the three primary controls.

An airplane that requires equal rudder, elevator, and aileron pressures for coordinated flight is called a *pilot's airplane*. Such airplanes are not common, but they do exist and they are a pleasure to fly.

Execution

Clear the surrounding airspace for traffic. Pick a point on the horizon and roll the airplane back and forth between left and right K-bank attitudes using a moderate rate of roll—not too fast, not too slow—while you keep the airplane's nose on the point. For the most part, your control inputs will be coordinated, but when you reach the K-bank attitude and reverse the roll, you must be very aggressive with the rudder in order to keep the nose on the point. If you are musically inclined and want a reference for timing, sing *Waltzing Matilda* while you perform the exercise.

CFI Notes

This exercise is not a Dutch Roll—a coupled oscillation in roll and yaw that is associated with swept-wing, jet aircraft. I therefore call it the control harmony exercise.

STEEP-BANK COORDINATION ROLLS

Purpose

To learn the proper steep-bank-turn entry and rollout procedures before you start practicing 360-degree steep turns. To develop attitude awareness.

Execution
Clear the surrounding airspace for traffic. Using a normal roll rate, roll back and forth between left and right K-bank attitudes. Entry: Make an audible power increase as you roll through 30 degrees of bank; increase the elevator back pressure when you reach 45 degrees of bank. Rollout: Remove the increased elevator back pressure and the previous power increase when you start the rollout. Remain in each steep-bank turn just long enough to ensure that you are maintaining altitude.

STALLS AND SLOW FLIGHT

ELEMENTARY AND SECONDARY STALLS

Purpose
To recognize a stall without using power or flaps. To recover from stalls with and without power, and to recognize secondary stalls. To develop attitude awareness.

Execution
Clear the surrounding airspace for traffic. Establish a power-off, best-glide descent, and then stall the airplane. To enter a stall, never continuously increase pitch attitude until the stall occurs. You should determine and establish the minimum pitch attitude that will result in a stall, and then maintain that attitude until the stall occurs. As airspeed decreases toward stall speed, you must continually apply more yoke back pressure in order to maintain the stall attitude. Performing stalls in this manner will increase your attitude awareness.

Recover with and without power by establishing the best-glide attitude. To experience a secondary stall, increase the pitch attitude immediately after you make the initial stall recovery.

Attitude
Set idle power and establish the C-pitch attitude and the B- or A-bank attitude. When ready to execute the stall, establish the stall attitude. Maintain bank attitude until the stall occurs. Do not let the nose drop below the H-pitch attitude during the recovery.

- No power recovery: Establish the H-pitch and B-bank attitudes. During one recovery, abruptly increase pitch to experience a secondary stall. Note the small effort that is required to make this happen. The secondary stall is really a low speed version of the accelerated stall.

- Power recovery: Establish the C-pitch and B-bank attitudes. Maintain this acceleration attitude until the controls come alive and you can establish the I-pitch attitude (best-rate climb) and climb back to the starting altitude. Again, during one recovery, abruptly increase pitch and notice how easily a secondary stall can be induced.

- Rudder usage: If the airplane rolls during a stall recovery, do not use aileron inputs. Stop the roll with rudder input. If you have difficulty with this concept or display a lack of confidence when you perform stalls, your instructor will have you practice the continuous stall exercise (refer to page 5 - 9), a coordination maneuver where the instructor holds the airplane in a power-off stall and you use the rudder to practice roll control.

In modern airplanes, ailerons can be used during recovery and doing so is a natural human reaction. However, this is a bad habit to develop. An unsolicited roll—a roll that was started by external forces, not you—requires immediate rudder input to prevent a spin from developing, to recover from a wake-vortex encounter, and to minimize yawing when flying in turbulence. **Rudder input must be your first reaction!**

When an unsolicited roll occurs, rudder input must be your first reaction.

If, due to poor judgment, a rapid roll occurs because you encountered a large aircraft's wake vortex, you must immediately follow full rudder input with full aileron input, decrease pitch attitude, and apply full power. Recovery is not guaranteed, but at least these actions will maximize your chance for survival.

If you desire, this is an excellent time for your instructor to demonstrate a spin. Keep your feet on the rudder pedals and your hand on the yoke, and follow your instructor's control inputs during the maneuver. Pay particular attention to the rudder's effectiveness during the spin recovery. Try one yourself. It's easy to do, and it will increase your self-confidence.

Power

Keep the power at idle for the power-off entries and recoveries. When you make a power-on recovery, simultaneously set full power and establish the C-pitch attitude.

Trim

Set for best glide.

Vision

Three points must be scanned. Evaluate these references while you scan the surrounding airspace for traffic.

- The wingtip and the horizon are used to establish pitch attitude and maintain bank attitude.

- The nose and the horizon are used to establish bank attitude and maintain pitch attitude.

- Stall entry: The pie-in-the-sky sight picture is used to supplement the first two references and evaluate roll, pitch, and yaw. Stall recovery: The Earth's surface and the horizon, framed by the nose cowl and the windshield's framework, are used to supplement the first two references and evaluate roll, pitch, and yaw.

CFI Notes

The proper visual references are critical, because pilots who watch the instruments during a stall are an accident waiting to happen. Make students aware of prestall control feel, airflow noise, and airframe buffet. To enhance this awareness, deactivate the stall warning horn—if possible—and cover the flight instruments. The resulting high level of awareness will make it much easier for you to teach landings. For stall warning horn deactivation, refer to 14 CFR part 91.213(d)(3ii and 4) and AC 91-67 Chapter 1, section 7b, and Chapter 2.

POWER-ON STALLS

Purpose

To execute and recover from a power-on stall in the takeoff or departure configuration. To recover with minimum loss of altitude. To develop attitude awareness.

Execution

Clear the surrounding airspace for traffic. Reduce power, select the takeoff flap setting if flaps are required, and establish a wings-level or a shallow-bank, best-rate climb attitude. When airspeed decreases below the best-rate climb speed, add full power and establish and maintain a pitch attitude that will cause a stall. Maintain the bank attitude until the stall occurs. *Note:* This is an expedited stall entry. There is no need to waste training time when you practice stalls.

Recover by simultaneously establishing the acceleration pitch attitude and a wings-level bank attitude. When control pressures increase, smoothly increase pitch to the best-rate climb attitude in order to minimize altitude loss. If the airplane rolls at the stall, correct with rudder, not aileron.

Attitude

When power is reduced, establish the B- or A-bank and the I-pitch attitudes. When full power is set, increase pitch to an attitude that will cause a stall, and maintain that attitude until the stall occurs.

To recover, simultaneously establish the C-pitch and B-bank attitudes for acceleration. When control pressures increase, smoothly increase pitch to the I-pitch attitude. Do not let the nose drop below the H-pitch attitude during the recovery.

Power

Reduce power in order to decelerate quickly. Set full power when the I-pitch attitude has been established and airspeed decreases below the best-rate climb speed.

Full power is normally used in training airplanes, but power is reduced to no less than 65 percent in high-performance airplanes (more than 200 horsepower) in order to prevent an excessive, nose-high pitch attitude during the stall.

Trim

Set for takeoff or initial climb.

Vision

Three points must be scanned. Evaluate these references while you scan the surrounding airspace for traffic:

- The wingtip and the horizon are used to establish pitch attitude and maintain bank attitude.

- The side of the nose and the horizon are used to establish bank attitude and maintain pitch attitude.

- Stall entry: The pie-in-the-sky sight picture is used to supplement the first two references and evaluate roll, pitch, and yaw. Stall recovery: The Earth's surface and the horizon, framed by the nose cowl and the windshield's framework, are used to supplement the first two references and evaluate roll, pitch, and yaw.

POWER-OFF STALLS

Purpose

To execute and recover from a power-off stall in the approach or landing configuration. To recover with minimum loss of altitude. To develop attitude awareness.

Execution

Clear the surrounding airspace for traffic. Set idle power and full flaps, and establish a wings-level or a shallow-bank, landing-approach attitude. When airspeed stabilizes, establish and maintain a pitch attitude that will induce a stall. Maintain the bank attitude until the stall occurs.

Recover by simultaneously establishing the acceleration pitch attitude, a wings-level bank attitude, and full power. When control pressures increase, select approach flaps, and smoothly increase pitch to the best-rate climb attitude in order to minimize altitude loss. Retract the remaining flaps. If the airplane rolls at the stall, correct with rudder, not aileron.

Attitude

With power at idle, establish the B- or A-bank and the H-pitch attitudes. When airspeed stabilizes, increase pitch to the stall attitude, and maintain that attitude until the stall occurs.

To recover, simultaneously establish the C-pitch and B-bank attitudes for acceleration. When control pressures increase, smoothly increase pitch to the I-pitch attitude. Do not let the nose drop below the H-pitch attitude during the recovery.

Power
Set idle power. When the stall occurs, set full power.

Trim
Set for the power-off approach configuration.

Vision
Three points must be scanned. Evaluate these references while you scan the surrounding airspace for traffic.

- The wingtip and the horizon are used to establish pitch attitude and maintain bank attitude.

- The nose and the horizon are used to establish bank attitude and maintain pitch attitude.

- Stall entry: The pie-in-the-sky sight picture is used to supplement the first two references and evaluate roll, pitch, and yaw. Stall recovery: The Earth's surface and the horizon, framed by the nose cowl and the windshield framework, are used to supplement the first two references and evaluate roll, pitch, and yaw.

SLOW FLIGHT

Purpose
To understand airplane control responses, performance, and aerodynamic characteristics when maneuvering at slow airspeed on the back side of the power curve.

Execution
Clear the surrounding airspace for traffic. Decelerate toward stall speed, extend the flaps, and maintain heading and altitude. As you decelerate, adjust pitch to maintain altitude, and as you approach the desired speed, add power to stop the deceleration. At the correct airspeed, the stall warning horn will activate and any significant increase in angle of attack will induce a prestall buffet.

During slow flight, you are flying on the back side of the power curve, a U-shaped curve that is based on horsepower and airspeed. The bottom of the curve is the minimum drag or minimum power-required speed. When you fly slower than the minimum power-required speed, you are flying on the back side of the power curve where power must be increased in order to maintain altitude, decrease airspeed, and overcome induced drag—drag generated by lift. When you fly faster than the minimum power-required speed, you are flying on the front side of the power curve where power must be increased in order to maintain altitude, increase airspeed, and overcome parasite drag—airframe drag.

You normally fly on the front side of the power curve, but your understanding of the back side of the power curve will reveal the cause of many takeoff accidents. When flying there with full power, you cannot accelerate without losing altitude. Many pilots have been killed because they attempted a takeoff when too heavy or when the density altitude was too high. Those airplanes became airborne in ground effect while on the back side of the power curve, and additional altitude or airspeed was impossible to obtain. Crash!

To recover from slow flight, use the go-around procedure: Set full power, approach flaps (slowly), and after you accelerate to climb speed, retract the remaining flaps. Maintain heading and altitude while accelerating to slow cruise or cruise flight.

Attitude

As the airplane decelerates, you will exceed the P-pitch attitude in order to maintain altitude.

Minimum airspeed can be maintained without using the airspeed indicator if you use the prestall buffet as a reference. If airspeed is correct, a slight, momentary pitch increase will induce the buffet. If you stare at the airspeed indicator, this maneuver is impossible to fly. Divide your attention at all times.

Elevator-throttle coordination must be used to maintain altitude and airspeed. If airspeed is high, increase pitch and decrease power slightly; as airspeed decreases, add power in order to maintain altitude. If airspeed is low, decrease pitch slightly and add power; as airspeed increases, decrease power slightly. If altitude is high, decrease pitch and power slightly; if altitude is low, increase pitch and power.

To recover, set full power and slowly lower the nose toward the C-pitch attitude in order to accelerate without losing altitude. During acceleration, pitch attitude controls altitude.

Power

To enter the maneuver and save training time, use idle power to decelerate rapidly. As the I-pitch attitude is approached and the controls start to feel mushy, increase power. The final power setting will be close to full power.

To recover, set full power. Do not reduce power until the flaps have been retracted and the airplane has accelerated to the desired cruise speed.

Trim

Rough-trim the elevator during the entry; fine-trim it once attitude and power are stabilized. Trimming decreases drag.

Vision

Three points must be scanned. Evaluate these references while you scan the surrounding airspace for traffic:

- The wingtip and the horizon are used to establish pitch attitude and maintain bank attitude.

- The side of the nose and the horizon are used to establish bank attitude and maintain pitch attitude.

- The pie-in-the-sky sight picture is used to supplement the first two references and evaluate roll, pitch, and yaw.

The cylinder head and engine oil temperature gauges must be scanned during this low-speed, high-power maneuver. If temperatures approach redline, terminate the maneuver and allow the engine to cool.

CFI Notes

Slow flight gives your student an opportunity to observe many important aerodynamic characteristics. Before performing the following demonstrations, you will improve student awareness if you deactivate the stall warning horn.

- Make turns, climbs, and descents while flying at minimum controllable airspeed. Demonstrate how you must continually evaluate elevator pressure and check for prestall buffet. Review how power and pitch attitude must work together when both airspeed and altitude must be controlled.

- Review the mushy control feel compared to normal- and high-speed control feel.

- Discuss torque effects and demonstrate why considerable right rudder must be used. Show that turning left is easier than turning right. Right rudder pressure is barely reduced to turn left, but increased considerably to turn right.

- Demonstrate adverse aileron yaw by making large aileron displacements.

- Review how turn rate varies with bank angle. Then, using a constant bank angle and increased power settings, demonstrate how turn rate varies with airspeed. To observe turn rate, watch the Earth's surface and the horizon. Do not use the turn coordinator.

- Discuss angle of attack as the airplane flies level, climbs, and descends. If students can visualize the relative wind in each of these cases, they will realize that prestall buffet always occurs at the same angle of attack. Increase pitch quickly to demonstrate that a stall depends on angle-of-attack, not airspeed. Point out the increased load factor that occurred during the rapid pitch increase.

- Increase speed slightly, and slowly roll into a steep bank attitude while maintaining altitude. When the prestall buffet occurs, stop rolling and note that stall speed increases as bank angle and load factor increase.

Then, while still experiencing the prestall buffet, add full power but maintain pitch attitude so that airspeed increases and the stall buffet stops. This proves that changing either pitch attitude or airspeed will change the angle of attack.

- Set full power, and note the low vertical speed that occurs when you attempt to climb with full flaps extended.

GROUND REFERENCE MANEUVERS

TRACKING OVER A ROAD

Purpose
To fly over a road at constant altitude. To recognize and compensate for wind drift. To control a linear ground track while properly dividing your attention.

Execution
Clear the surrounding airspace for traffic. Establish slow cruise and fly over a road at 600 to 1,000 feet above ground level (agl) when a crosswind is present. To do this, you must point the airplane's nose into the wind. The difference between the airplane's heading and its ground track is the wind correction or drift angle. Reverse course and apply the wind correction angle in the opposite direction.

Vision
Use the nose and wingtip references to maintain attitude. Continually cross-check airspeed, altitude, the road, and the surrounding airspace for traffic.

TRACKING PARALLEL TO A ROAD

Purpose
To fly parallel to a road at constant altitude. To recognize and compensate for wind drift. To plan and control a linear ground track while properly dividing your attention.

Execution
Clear the surrounding airspace for traffic. Establish slow cruise and fly parallel to a road, offset a suitable distance, at 600 to 1,000 feet agl when a crosswind is present. Visualize the desired ground track on the Earth's surface, and pick out specific points that you must overfly in order remain parallel to the road.

The secret for performing ground reference maneuvers is to visualize the desired ground track and then overfly it.

After you turn parallel to the road, some instructors will tell you to note its position in relation to the wing or the wing strut, and then turn as necessary to keep the road in that same relative position. Yes, that works in this case, but it does not work with the advanced ground reference maneuvers or the traffic pattern's crosswind and base legs. Learn to visualize your desired ground track on the Earth's surface and pick out specific points that must be overflown.

Establish the wind correction angle that is required in order to remain over the desired ground track. Reverse course and again apply the wind correction angle and the ground track planning requirements.

Vision
Use the nose and wingtip references to maintain attitude. Pick out surface references that represent the desired ground track. Continually cross-check airspeed, altitude, the desired ground track, and the surrounding airspace for traffic.

RECTANGULAR COURSE

Purpose
To fly a rectangular pattern while maintaining a constant distance from a rectangular reference and a constant altitude. To recognize and compensate for wind drift. To plan and control the ground track while properly dividing your attention.

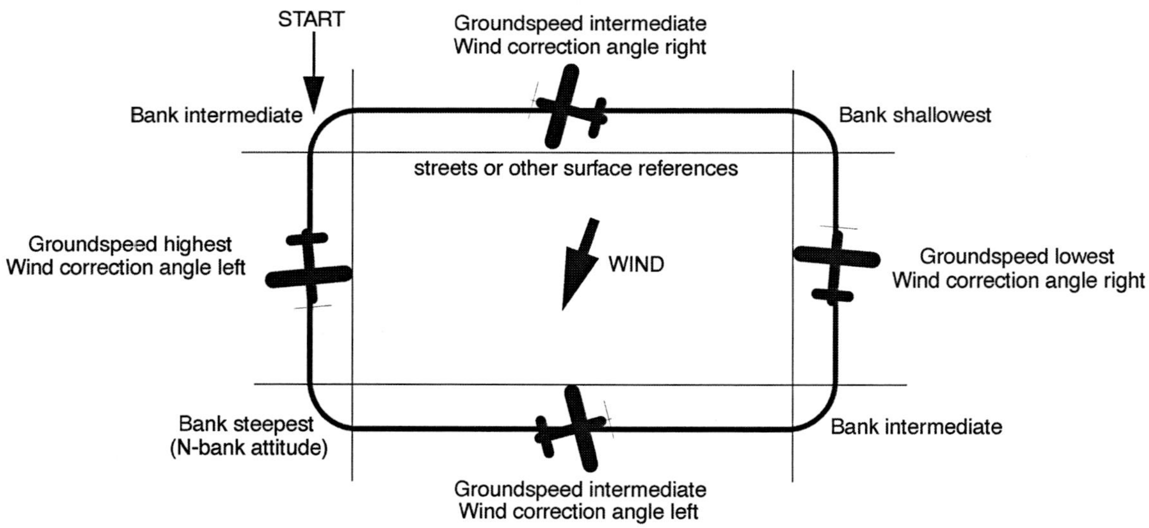

Rectangular Course

Execution
Clear the surrounding airspace for traffic. Establish slow cruise, pick a rectangular pattern—usually four roads—and

fly around it starting on a downwind heading. Remain parallel to each side, offset a suitable distance, at 600 to 1,000 feet agl. To quickly determine wind correction angles and remain equidistant from each side of the rectangle, you must pick points on the desired ground track and overfly them.

To maintain a constant radius when turning from one leg to the next, your bank angle must be adjusted for groundspeed. The steepest bank angle occurs when groundspeed is greatest; the shallowest bank angle occurs when groundspeed is least. You start the maneuver on a downwind heading so that your first turn requires the steepest bank, which for this maneuver should be a medium bank—the N-bank attitude. Airspeed remains constant except for a slight decrease that occurs during turns due to the increased drag.

Vision

Use the nose and wingtip references to maintain attitude. Pick out surface references that represent the desired ground track. Continually cross-check airspeed, altitude, the desired ground track, and the surrounding airspace for traffic.

S-TURNS

Purpose

To fly a series of constant-altitude, constant-radius semicircles that lie on opposite sides of a straight reference line such as a road when a crosswind is present. To recognize and compensate for wind drift by varying bank angle. To plan and control a circular ground track while properly dividing your attention. S-turns are an orientation maneuver, not a ground track maneuver.

Execution

Clear the surrounding airspace for traffic. Start upwind of the road and fly downwind, perpendicular to it, at 600 to 1,000 feet agl. The instant you cross the road, start the first 180-degree turn with a medium bank angle.

For the remainder of the maneuver, the airplane never flies straight and level. Each time the road is crossed, the airplane must be perpendicular to the road and rolling through a wings-level attitude as the direction of turn is reversed.

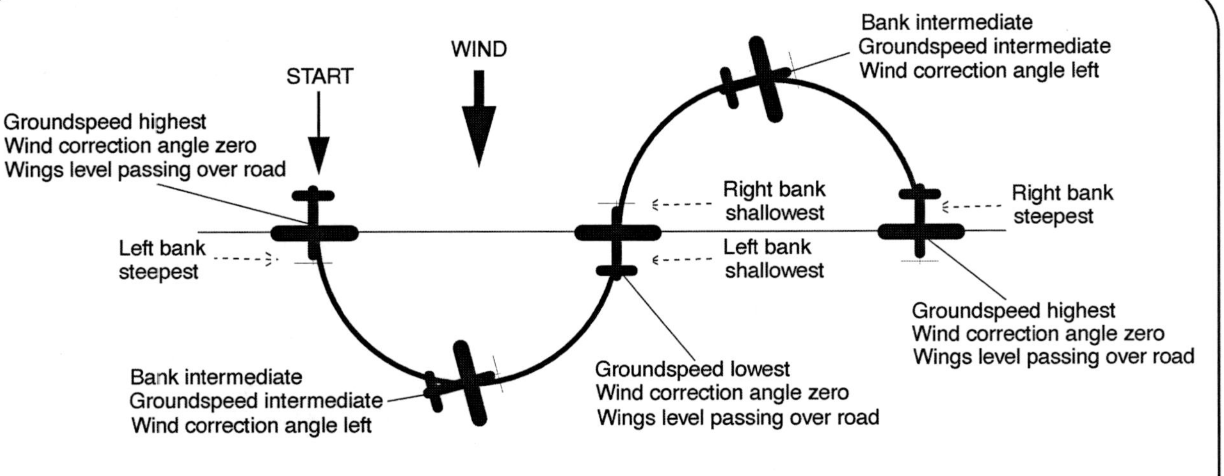

S-Turns

Attitude

Pitch attitude and airspeed remain constant throughout the maneuver. Bank attitude and groundspeed continually vary. The steepest bank occurs when you are flying downwind—at the point of greatest groundspeed. The shallowest bank occurs when you are flying upwind—at the point of slowest groundspeed.

The following occurs at the first three road crossings:

First crossing The airplane is flying downwind at the fastest groundspeed, and you establish the N-bank attitude. During this turn, groundspeed will slowly decrease. As you complete the first 90 degrees of this turn, it will become apparent that to be perpendicular to the road when you cross it, you must decrease the bank angle. The amount of change depends on wind velocity.

Second crossing Now the airplane is flying upwind at the slowest groundspeed. As the road is crossed, you roll the airplane from a shallow left bank to a shallow right bank. Make certain that the second shallow bank is no steeper than the first. As you complete the first 90 degrees of this turn, it will become apparent that to be perpendicular to the road when you cross it, you must increase the bank angle back to the N-bank attitude.

Third crossing Like the first crossing, the airplane is flying downwind at the fastest groundspeed. As the road is crossed, the airplane is rolled from a right N-bank attitude to a left N-bank attitude. Repeat the maneuver as necessary.

Power
Slow cruise.

Trim
Trimmed for slow cruise flight.

Vision
Use the nose and wingtip references to maintain attitude. Continually cross-check airspeed, altitude, the road—*not ground track*—and the surrounding airspace for traffic.

S-turns are an orientation maneuver, not a ground track maneuver.

The semicircles should be equal in size, but ignore that terribly misleading requirement. Orientation is your primary concern, not ground track. Here's the secret: Concentrate on being perpendicular to the road when you cross it, and on making the second shallow bank no steeper than the first when you cross the road and roll from one shallow bank to the next. If you do this, the semicircles will have equal radii.

In the previous illustration notice the left wind-correction angles that occur after 90 degrees of each turn. When you cross-check the road at this point during the first turn, it will be obvious that bank angle must be reduced. When you cross-check the road at this point during the second turn, it will be obvious that bank angle must be increased.

CFI Notes
Ground reference maneuvers other than steep spirals—a commercial pilot requirement—are started with a downwind entry, so that the maneuver's highest groundspeed and steepest bank angle will occur at this point. During the rest of the maneuver, groundspeed and bank angle will be equal to or less than that used for the entry.

If you start a ground reference maneuver by flying upwind, an unobservant or improperly trained student may use too steep a bank when downwind or may make an excessive pitch increase upon sensing the increased groundspeed.

Make certain that students watch the nose reference when rolling between turns as they cross the road. To do otherwise makes it impossible to maintain altitude.

Students are often mislead by the term steepest bank angle. This does not mean that a steep bank angle—the K-bank attitude—is used during the maneuver.

Lift is a powerful force and small bank angle changes produce noticeable results.

Students often make excessive bank angle changes when they try to correct ground track during ground reference maneuvers. Lift is a powerful force and small bank angle changes produce noticeable results in light to moderate winds. Students should make a reasonable correction and then take a moment to evaluate the result before they make another correction.

SIDESLIPS

Purpose

To move laterally through the air while you maintain a constant heading so that wind drift can be nullified during crosswind approaches and landings. To fly the airplane with crossed controls.

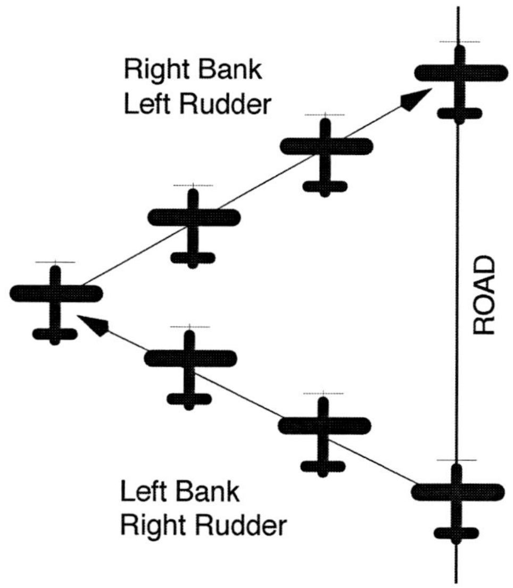

Sideslipping Away From and Toward a Road

Execution
Clear the surrounding airspace for traffic. Fly over a road at 1,500 feet agl when a crosswind is present and start a power-on or power-off landing descent. Bank the airplane to the left with aileron, keep the airplane parallel to the road with rudder, and maintain pitch attitude with elevator. The airplane will move laterally away from the road.

Do not let the nose of the airplane point toward or away from the road. Rudder must be used independently of the ailerons to keep the airplane parallel to the road at all times. Reverse the bank angle and move laterally back toward the road.

If bank is increased excessively, you can no longer maintain heading with rudder. Pilot's call this *running out of rudder*, and that's what limits your ability to land in a crosswind.

Finally, start another descent over the road, but sideslip so that you remain over the road's centerline. The controls will be crossed—right rudder and left aileron or left rudder and right aileron—but don't think about this. Just use rudder to keep the airplane parallel with the road's centerline and use bank angle to remain over the centerline. It's that simple.

Wind usually turns counterclockwise and decreases in velocity as you descend.

Note: In the northern hemisphere, wind velocity usually decreases and *backs* (turns counterclockwise) as you descend, reducing the required crosswind correction. As you climb, wind velocity normally increases and *veers* (turns clockwise), increasing the required crosswind correction.

Attitude
Maintain the C- or H-pitch attitude. Rudder controls yaw to stay parallel with the road. Ailerons controls bank angle in order to stay over or move laterally away from or toward the road. Elevator controls pitch attitude and airspeed.

Power
Approach power if you use the C-pitch attitude; idle power if you use the H-pitch attitude.

Trim
Trim for the power-on or power-off landing descent. Do not trim during the slip.

Vision

Three points must be scanned. Evaluate these references while you scan the surrounding airspace for traffic:

- The wingtip and the horizon are used to establish pitch attitude and maintain bank attitude.

- The nose and the horizon are used to establish bank attitude and maintain pitch attitude.

- The Earth's surface and the horizon, framed by the nose cowl and the windshield's framework, are used to supplement the first two references and evaluate roll, pitch, yaw, road (runway) alignment, and lateral motion.

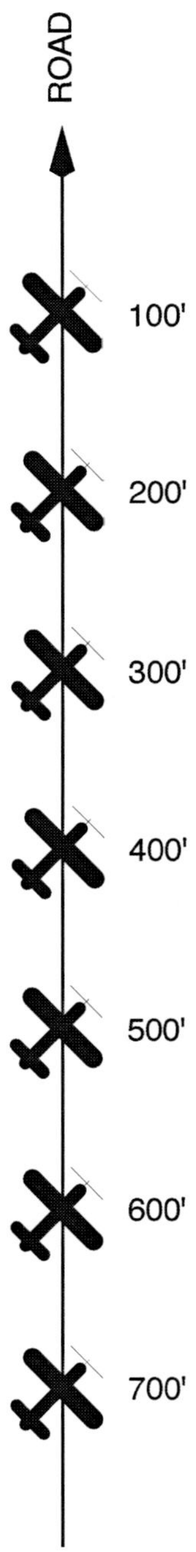

Forward Slips

FORWARD SLIPS

Purpose

To rapidly lose altitude by maximizing drag during a power-off descent in order to correct your glidepath when landing. To coordinate the lift and thrust vectors in order to maintain the desired ground track. This is strictly an attitude-only maneuver, because indicated airspeed will be incorrect due to abnormal airflow at the pitot tube and static ports.

Execution

Clear the surrounding airspace for traffic. Fly down a road at 1,500 feet agl and begin a power-off descent. For a maximum performance forward slip, simultaneously apply full right or left rudder and enough opposite aileron so that the wing's lift vector keeps you over the road—yes, you use bank angle to steer the airplane, not rudder. If only a small glidepath correction is required, you can make a gentle forward slip by using partial rudder and a shallow to medium bank. To terminate the slip, simultaneously release the flight controls and the airplane will automatically return to a normal attitude.

Check your POH before you perform a forward slip with flaps extended. If it says "not recommended," that does not mean prohibited. If you were making a forced landing with full flaps extended and you were high, the forward slip is your only option for a safe landing—but you must ignore indicated airspeed and never let the nose rise above the H-pitch attitude.

The descent rate can be increased considerably if you lower the nose well below the H-pitch attitude, but during the recovery you must aggressively restore the H-pitch attitude in order to prevent an airspeed increase. Excessive airspeed on final approach during a forced landing is just asking for trouble.

If you question an airplane's ability to execute a maximum performance forward slip with full flaps extended, try it at altitude. If full flaps cause the horizontal stabilizer to stall, the airplane will pitch down somewhat until the horizontal stabilizer regains normal airflow, and then the airplane will pitch up to the previous attitude. To stop these oscillations, increase the nose-down pitch attitude when slipping.

When a crosswind exists, many instructors will tell you to bank into the wind when executing a forward slip so that it is easy to transition to a sideslip for landing. That's fine if you start the forward slip on final approach. However, if you are starting your turn from base leg onto final approach or if you are in that turn when you recognize that you are too high, you must apply *top rudder* so that you can forward slip while turning, regardless of the wind conditions. Top rudder means the right rudder when in a left turn and left rudder when in a right turn.

Always be skeptical of reported winds. What you observe is what you have.

When you terminate the forward slip, remain aligned with the runway for a moment, evaluate actual wind conditions, and then sideslip as necessary for landing. Always be skeptical of reported winds. Base your corrections on airplane movement in relation to the runway's extended centerline.

Be aggressive when you need to use a forward slip. By doing this, your correction will occur while you are at altitude so that this type of maneuvering will not be required when you are close to the ground.

Attitude

Use elevator to keep the nose at or below the H-pitch attitude and use full rudder for maximum performance. This is an pitch-attitude maneuver, not an airspeed maneuver. Use ailerons and bank angle to steer the airplane.

Power

Idle for maximum performance.

Trim
Trim for the power-off, landing descent. Do not trim during the slip.

Vision
Three points must be scanned. Evaluate these references while you scan the surrounding airspace for traffic:

- The wingtip and the horizon are used to establish pitch attitude and maintain bank attitude.

- The nose and the horizon are used to establish bank attitude and maintain pitch attitude.

- The Earth's surface and the horizon, framed by the nose cowl and the windshield's framework, are used to supplement the first two references and evaluate roll, pitch, yaw, and ground track.

CFI Notes
Sideslips and forward slips can be confusing to students because the controls are used in the same manner. Explain that in both cases the ailerons (bank angle) are used to control ground track. In a sideslip partial rudder is used to keep the airplane parallel with the runway centerline. In a forward slip, full rudder is used if maximum performance is required. Yes, students can use partial rudder in a forward slip and be less aggressive, but they are better off to make an aggressive altitude correction while at altitude and avoid this type of maneuvering when they are close to the ground.

REVIEW TEST

1. Describe the five senses and their application to flying.
2. Describe airplane attitude and the six types of motion.
3. List the P-I-T-C-H and B-A-N-K attitudes, performance objectives, and power settings.
4. How are pitch and bank attitudes established and maintained?
5. How do you neutralize the flight controls?
6. Describe the two-step, control-input rule.
7. Describe the two-step correction rule.
8. Describe the term "indexing the throttle."
9. How do you use elevator trim?
10. How do you use rudder trim?
11. What happens if you attempt to fly the airplane with the trim controls?
12. How do you perform hands-off flight and why?
13. How should you adjust the pilot's seat?
14. What is slow cruise and when is it used?
15. Before making any turns and during prolonged climbs and descents, you must?
16. During turns, what is the proper way to determine correct rudder usage?
17. Describe the proper climb entry and level-off procedures.
18. Describe the proper descent entry and level-off procedures.
19. What is the primary purpose of the cruise climb and the cruise descent?
20. What is a best-rate climb?
21. Describe the constant-airspeed and constant-altitude coordination exercises.
22. What is the purpose of the control-harmony exercise?
23. What two control inputs reduce the angle of attack during stall recoveries?
24. What should your first reaction be if an airplane starts an unsolicited roll?
25. How are spins avoided during stall recoveries?
26. You are doing 360-degree turns in a 20-knot wind. What will be the airspeed change?
27. How do you compensate for wind drift during turns and straight flight?
28. What is the purpose of the rectangular course?
29. What is the purpose of S-turns?
30. What is the purpose of a sideslip?
31. What is the purpose of a forward slip?
32. Why do you always start ground reference maneuvers while flying downwind?

ATTITUDE PLUS POWER EQUALS PERFORMANCE—THAT'S THE CRITICAL PREREQUISITE FOR COMPETENT AIRMANSHIP. IF YOU DON'T FORCE YOURSELF TO USE THE PROPER VISUAL REFERENCES AND TO CONTINUALLY DIVIDE YOUR ATTENTION, YOU WILL NEVER MASTER THE ART OF FLYING.

CHAPTER 4

NORMAL TAKEOFFS AND LANDINGS

People place the wrong emphasis on takeoffs and landings. Passengers judge pilots by these events, yet they have no idea of whether or not the flight was flown safely; beginning student pilots think about landings almost continuously—a distraction that hampers their ability to learn the prerequisite elements of instruction.

The foremost rule is to stay relaxed and use the proper visual references. Learning to land is frustrating because you must continually evaluate attitude, altitude, runway orientation, and six types of motion while you operate a machine that has three control axes and a throttle. The pressure is on, because if you goof up, the result can be disastrous. Fortunately, your flight instructor will not allow that to happen.

When you solo, your first words will be, "That wasn't too hard."

The first step is to have patience and realize that you must develop new habit patterns that were never needed during your earthbound endeavors. The training time required to master landings and fly solo depends on how fast you can change your earthbound habits, many of which have no place in an airplane. When you do solo—a moment you'll never forget—your first words will be, "That wasn't too hard."

TAKEOFFS

NORMAL TAKEOFF AND THE DEPARTURE PROFILE

Purpose

To reach a safe altitude as fast as possible. To minimize your vulnerability during the most dangerous portion of any flight—the takeoff and initial climb.

That danger exists because of high internal engine stresses, limited pilot options, reduced performance due to aircraft weight, and the poor crashworthiness of airplanes that can easily result in ruptured fuel tanks and fire.

Safe altitude is the lowest altitude that provides a suitable emergency landing site.

Safe altitude is based on pilot judgment, and it represents the lowest altitude that provides a suitable emergency landing site in the event of an engine failure. In most cases, 1,000 feet above the departure airport meets this requirement.

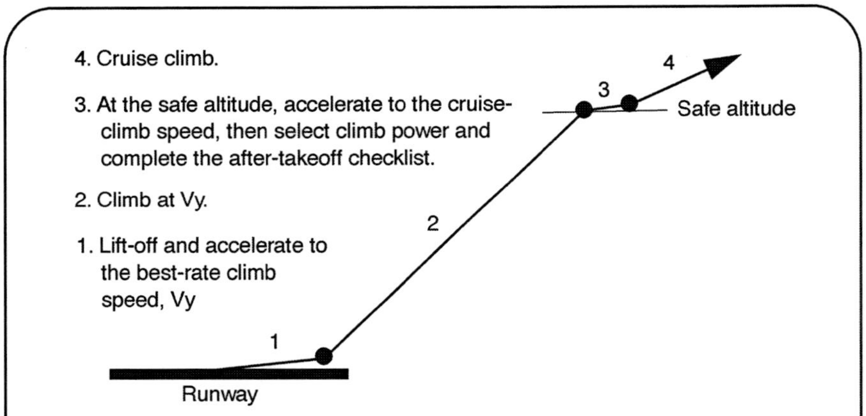

The Departure Profile

Execution

While taxiing to the runway, monitor the airport's windsock and check the magnetic compass for accuracy. When cleared onto the runway, check the final approach for aircraft, and perform the lights, camera, action mental checklist. Set takeoff power, check engine rpm and oil pressure, and maintain a level pitch attitude during initial acceleration. When the elevator becomes effective, establish the takeoff pitch attitude and *maintain that attitude* after liftoff in order to quickly accelerate to the best-rate climb speed (V_Y).

At Vy, establish the best-rate climb attitude and maintain it until you reach the safe altitude. Now, if you are departing the area, establish the cruise-climb attitude, accelerate to the cruise-climb speed, and set climb power. If you are remaining in the traffic pattern, establish the slow-cruise attitude at pattern altitude, accelerate to the slow-cruise speed, and set slow-cruise power.

Attitude

At the start of the takeoff roll, maintain the C-pitch attitude for initial acceleration. When elevator inputs become effective, establish the T-pitch attitude and *maintain it* while the airplane accelerates, lifts off, and accelerates to V_Y.

Learn to feel control pressures and listen to engine noise.

At Vy establish the I-pitch attitude and maintain it until you reach the safe altitude. Now, re-establish the T-pitch attitude for either the cruise climb or slow cruise. You must learn to sense the approach of V_Y by feeling control pressures and listening to engine noise. Control pressures and engine noise increase as airspeed increases.

You can determine the T-pitch attitude while the airplane is on the ground.

You can determine the T-pitch attitude on the ground providing the POH does not prohibit the following procedure. While you sit in the pilot's seat, your instructor will put the airplane in the T-pitch attitude by pushing down on a fuselage bulkhead that's just forward of the vertical stabilizer. Now, look outside and mentally note the nose and wing position in relation to the imaginary horizon.

To maintain the T-pitch attitude at liftoff for acceleration, you must immediately move the yoke forward slightly.

To maintain the T-pitch attitude at liftoff and accelerate rapidly to V_Y, you must *immediately* move the yoke forward slightly so that the pitch attitude does not increase. It wants to increase because the horizontal stabilizer's moment arm, down force, and airspeed all increase at this time.

The moment arm increases, because its forward end moves from the point of main-wheel ground contact to the airplane's center of gravity. The down force increases, because the air flow changes as you leave ground effect. (This does not occur with T-tailed airplanes because the horizontal stabilizer is not affected by the airflow behind the wing.) The increasing airspeed pitches the nose up, because poorly trained pilots will hold the elevator in a fixed position.

Takeoffs in overloaded airplanes or at high density altitudes have caused many fatal accidents.

Ground effect is an aerodynamic change that occurs while the airplane's altitude is less than half of its wingspan above the runway. Due to the extremely low altitude, the wing cannot deflect the relative wind downward in the normal manner, so induced drag—the drag that lift creates—decreases. This drag reduction increases performance. Attempting to fly an overloaded airplane or fly at high density altitudes has caused many fatal accidents, because the airplane would lift-off in ground effect but could not climb above it.

When you enter or leave ground effect, the airflow behind the wing changes—as it does with flap extension or retraction—and that changes a conventional horizontal stabilizer's angle of attack. Pitch attitude increases when you climb out of ground effect during a takeoff; it decreases when you descend into ground effect during a landing.

Power

Never reduce power prior to reaching the safe altitude and the desired climb speed.

Apply full power and check rpm and oil pressure. This is your first opportunity to check the engine's maximum power output, and it gives you ample opportunity to abort the takeoff if the indications are abnormal. Full power, static run-ups are not made prior to takeoff in modern airplanes because of insufficient engine cooling and propeller abrasion. Never reduce engine power until you reach the safe altitude and accelerate to your desired airspeed.

Pay attention to the following terms in the airplane's POH:

Takeoff power A time-limited power setting. Example: 2,700 rpm, limited to five minutes. This power setting does not apply to basic training airplanes and is not published in the POH.

Maximum continuous power can be used forever.

Maximum continuous power A power setting that is not time limited—you can use it until you run out of gas. For basic training airplanes, maximum continuous power is used for takeoff and all climbs. This power setting requires full throttle in basic training airplanes, so I use the term *full power* in this manual.

Climb power The cruise-climb power setting. If the POH specifies full throttle, that procedure is critical, because the excess fuel is used to cool the engine's exhaust valves.

Improperly trained pilots often make an early power reduction after takeoff because they hear the engine rpm increase to a high value or they want to "save" their engine. The high rpm occurs because airspeed is too fast. If they would climb at V_Y, the engine would sound normal and create less airport noise. Saving an engine in this case is a myth, and these pilots are compromising their safety by operating contrary to procedures that are sanctioned by all engine manufacturers.

Trim

As specified in the POH. If not specified, set neutral trim.

Vision

A student who ignores wind indicators before takeoff has a serious, disqualifying awareness problem.

Proper vision starts while taxiing to the runway. You must look for a windsock or other wind indicator to determine the actual wind. Never rely solely on reported winds. A student who ignores wind indicators has a serious, disqualifying awareness problem.

During the takeoff roll, look straight down the runway. At liftoff, however, you must immediately shift your eyes to the pie-in-the-sky sight picture so that you can prevent the pitch attitude from increasing until you have accelerated to V_Y.

CFI Notes

Do not teach students to hold an airplane on the ground and "rotate it" into the air at a specific airspeed. When light airplanes, T-tailed or not, are flown in this manner, weight transfers onto the nose gear while accelerating, a grossly incorrect procedure that causes wheelbarrowing.

No competent pilot would ever trust single-source data without confirmation, and light airplanes have only one airspeed indicator. T-tailed general-aviation airplanes with their lack of slow-speed elevator authority prompted manufacturers to use the concept of "rotation speed," which is only valid in large airplanes that have two independent airspeed indicators.

You can trust the wing, but not the airspeed indicator.

Regardless of tail configuration, teach attitude awareness and teach students to establish the T-pitch attitude when the elevator becomes effective. With this attitude, weight distribution is correct, and the airplane will fly itself off the runway when the wing is ready to fly. You can trust the wing!

If you fly a T-tailed airplane, apply aft-elevator pressure when the airplane starts accelerating. However, the instant the nose starts to rise, *immediately* remove most of that input. The airplane will then hold attitude and perform just like airplanes that have a conventional horizontal stabilizer.

FLYING THE TRAFFIC PATTERN

Purpose

To fly a rectangular ground track that is oriented to a runway. To recognize and compensate for wind drift and traffic separation.

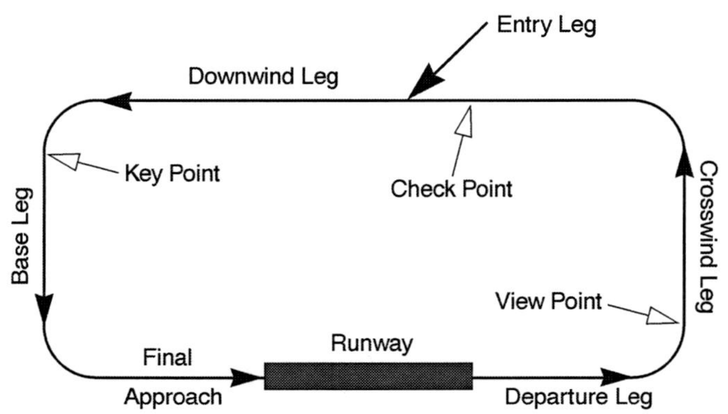

Airport Traffic Pattern

Execution

Clear the surrounding airspace for traffic. After takeoff, track the extended runway centerline until 300 feet below pattern altitude, but before you turn onto the crosswind leg, clear your airspace and use the wing just like you would use a pool cue to lineup your shot—the crosswind leg's ground track in this case. I call this *pool cue flying*, and you should do it before you turn onto any one of the traffic pattern's legs. Now, if you keep the airplane on the selected ground track, you will fly a rectangular traffic pattern.

Use the wing just like you would use a pool cue to lineup your shot.

If your initial climb out requires a wind correction angle, look ahead of or behind the wing by that amount when you pick the new ground track. The legs of the traffic pattern must be either perpendicular or parallel to the runway.

The view point is located at the beginning of the crosswind leg.

After you turn onto the crosswind leg and level the wings, you are located at the *view point* that's depicted in the previous illustration. Carefully check the entry and downwind legs for traffic, and then turn onto the downwind leg.

The check point is located on the downwind leg abeam the departure end of the runway.

At the *check point*, located abeam the departure end of the runway, apply carburetor heat if applicable, check the engine's oil pressure and oil temperature gauges, review the GUMPS mental checklist, call the tower—always the last priority—and carefully check for traffic that may be ahead of you on the downwind leg, base leg, or final approach.

If the tower says that you are number three for landing, there is a trick for quickly locating the preceding traffic—immediately look at the runway threshold and work backward. The number one airplane will probably be on final, the number two airplane will probably be on base, and you are next.

The starting point for your descent to the runway will depend on one of two traffic situations:

The key point is located at the beginning of the base leg.

Number one to land Start the descent when opposite your intended touchdown point on the runway. Turn onto the base leg when the touchdown point is 45 degrees behind the wing. Upon completion of this turn, you are located at the *key point*, the point where the landing evaluation process begins as described in the next section. Visually extend the runway's centerline on the ground for reference, and plan your turn to final so that you intercept that line.

Following other traffic Remain at pattern altitude until you are in position for a normal descent to the runway. Descending too early eliminates forced landing options. The amount of traffic will dictate whether you start your descent on base leg or on final approach. Always start the base leg turn just after the airplane you are following is on final approach and passes abeam your wingtip.

Attitude

The T-pitch attitude is used to fly the downwind leg. The C-pitch attitude is used for a power-on approach descent. The H-pitch attitude is used for a power-off approach descent. The N-bank attitude is normally used for all turns.

Power
Climb: full power. Level flight: slow-cruise power. Power-on descent: landing-approach power—1,300 to 1,400 rpm. Power-off descent: idle power.

Trim
As required.

Vision
Cross-check the airplane's attitude, ground track reference points, and wind correction angle. Use points on the ground to define the desired ground track. If you fly over these points, you will automatically establish the proper wind correction angle.

Use points on the ground to define the planned ground track.

CFI Notes
Do not assume that training time can be saved by teaching the objectives of elementary ground reference maneuvers in the traffic pattern. Training time will, in fact, be increased as will your student's learning-to-land frustration level. Students must master ground reference maneuvers before they are exposed to the variations and distractions of flying the airport traffic pattern and learning to land.

CONSTANT ATTITUDE AND POWER, VARIABLE AIRSPEED APPROACH

General
I read about this approach years ago in a magazine article that was written by Wolfgang Langewiesche, the author of *Stick and Rudder.* A few years later, much to my surprise, I was taught to fly it when I went through U.S. Army flight training. I have used it and taught it ever since, because it fully complies with my flight training philosophy, as explained in the introduction to this manual. I use the acronym CAAPVAA (pronounced cap-va) to reference this procedure.

There are several ways to make approaches, but none provide the benefits of CAAPVAA: Continual evaluation of all approach variables, confident forced landings, safe night landings, and the continual self-enhancement of your approach proficiency for as long as you fly airplanes.

The airplane slowly settles toward the runway. It does not dive toward the runway.

CAAPVAA is easy to perform, easy to learn, and it is the prerequisite for consistently good approaches and landings. With this procedure the airplane slowly settles toward the runway. It does not dive toward the runway. That's the difference.

Purpose

To transition from pattern altitude to the landing flare point. To master a procedure that generates consistently good landings and continually enhances pilot proficiency.

Execution

Clear the surrounding airspace for traffic. Extend the first increment of flaps and establish the landing attitude and power setting to start the approach. Ideally, you control flap extension and ground track so that attitude and power changes are not required until the landing flare. CAAPVAA is used for both power-on and power-off approaches.

Power-on approach Extend the first increment of flaps, reduce power to approximately 1,300 to 1,400 rpm—note the engine sound and use that as a reference just like you did for the slow-cruise descent—and establish the C-pitch attitude. As you extend additional flaps on the base and final legs, drag increases and airspeed decreases. After you extend full flaps, airspeed will be close to the required final approach speed, but a slight pitch adjustment may be required to obtain the exact speed.

If you use the proper pitch attitude, airspeed will decrease to the required final approach speed when full flaps are extended.

Power-off approach Extend the first increment of flaps, reduce power to idle, and establish the H-pitch attitude. After you extend full flaps, airspeed will be close to the required speed, but a slight pitch adjustment may be required to obtain the exact speed.

During either approach, your goal is to reach the landing flare point at the correct airspeed without changing attitude or power. No one can do this on every approach. If you get high, you must reduce power, widen-out your traffic pattern, or perform a forward slip. If you get low, you must add power. That's okay, just keep on trying. The CAAPVAA approach will automatically increase your insights and develop a high level of awareness and proficiency. Use the following tools to accomplish that important goal:

The turn to base leg If number one for landing with a calm to light wind, you make this turn when the runway threshold is approximately 45 degrees behind the wing. Start the turn earlier in strong winds. Upon completion of the turn, you are at the key point where you start evaluating your descent angle.

Evaluating the descent angle You do this continually while on the base and final legs. In the following illustration, A is the distance from your position to the runway threshold; B is the distance from the departure end of the runway to the horizon, which is always located at the base of mountains.

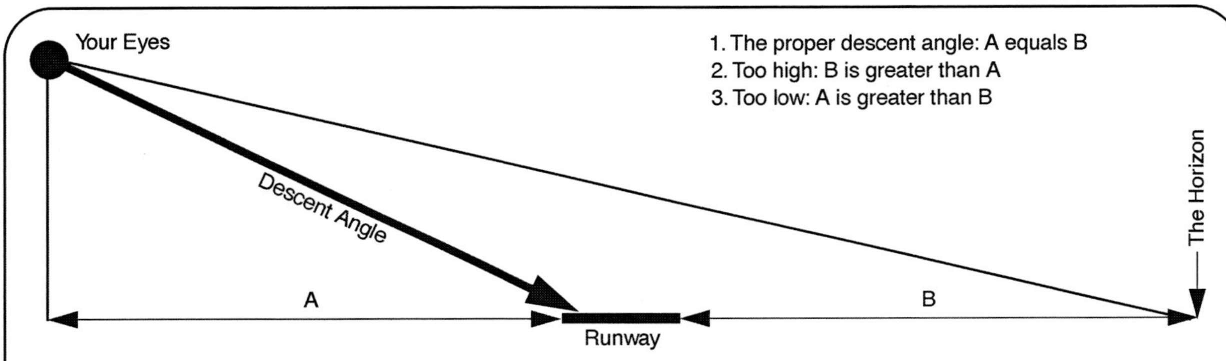

Evaluating The Descent Angle

At the key point, A will be shorter than B, but as you continue the descent, equalization will occur. Now you must use good judgment with respect to the wind gradient and updrafts or downdrafts (also called risers or sinkers):

- If A continues to equal B, your glidepath is correct.

- If you get too low—A is greater than B—correct your glidepath by shortening your ground track to the runway or by delaying flap extension in order to keep your speed up. Remember, wind velocity normally decreases and groundspeed increases as you approach the runway.

- If you get too high—B is greater than A—correct your glidepath by lengthening your ground track or by extending flaps earlier than normal in order to reduce airspeed. Maintain your pitch attitude. If you decrease pitch attitude and dive toward the runway, airspeed will increase and you will float during the landing.

Base leg orientation Normally, your base leg is flown perpendicular to the final approach, but if you are high or low, you can turn away from or toward the runway in order to increase or decrease your ground track.

Base leg orientation, flap selection, and the turn to final are your tools for modifying the ground track and descent angle.

Flap selection Lift and drag increase as flaps are extended, and the increased drag decreases airspeed and steepens the descent. If your airspeed on the downwind leg is 85 knots, it may be 80 knots after you begin the descent and extend the first increment of flaps, 75 knots with two increments of flap extension, and 65 knots, the final approach speed, with full flap extension.

Traditionally, the second and third (final) increments of flaps are selected on base leg and final approach, respectively, but this is not a rigid rule. Flap extension must be based on your judgment of descent angle, ground track, wind, updrafts and downdrafts.

You must anticipate the need for flap extension, because when you are on the proper glidepath with a normal wind gradient, you are about to go high. So when everything looks correct, extend the next increment of flaps. If you delay flap extension until you are too high, a power reduction will be required.

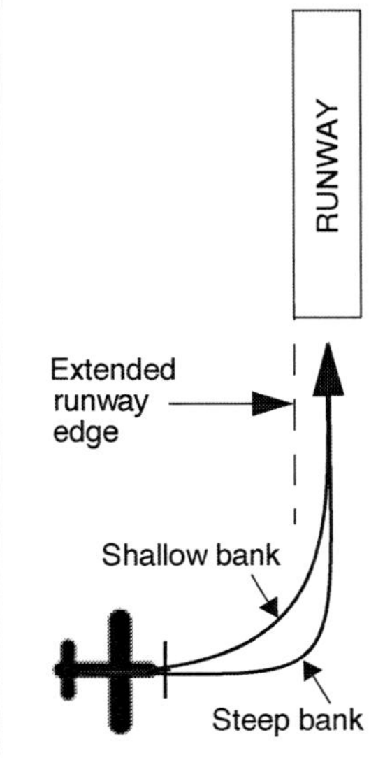

Ground Track and Bank Angle

The turn onto final approach Make a medium-bank turn, but if you are too high or too low, adjust the descent rate and ground track with bank angle as depicted on the left. Ground track and descent rate will both increase if you make a steep-bank turn onto final approach; they will both decrease if you make a shallow-bank turn.

Here is a trick for intercepting the extended runway centerline when you turn onto final approach: Ignore the centerline and rollout on the extended runway edge as depicted on the left. Most pilots overshoot the extended runway edge and wind up on the runway centerline. Look, your instructor can smile.

Anticipating updrafts and downdrafts The Sun's radiation heats the ground, not the air. The ground, in turn, heats the air by conduction. When the proper conditions exist, hot air will rise causing updrafts—also called thermals.

High temperatures are most likely to occur when the Sun's rays are vertical to the surface in question and when that surface is smooth, dark in color, and dry—smooth surfaces get hotter than rough surfaces. Any surface that contains moisture—plants, trees, or water—keeps warm air near the ground because evaporation slows the heating process.

Wind normally backs and velocity decreases as you descend toward the ground.

Anticipating wind gradient changes As the airplane descends toward the surface, wind direction normally *backs* (moves counterclockwise) and velocity decreases. If the reported surface wind is 040 degrees at 8 knots, the wind at traffic pattern altitude could easily be 060 degrees at 15 knots. As the headwind component decreases, groundspeed increases, and you will go above the desired glidepath if a correction is not made.

Success! You approach the flare point and you have not changed attitude or power during the approach. Congratulations, you won the contest and are on your way to becoming a pilot, not an airplane driver. There's a big difference.

As I stated earlier, this will not occur on every approach. If you get too high you must reduce power; if too low, you must add power. If the power changes are slight, you can maintain pitch attitude, but If the power changes are significant, you must change pitch attitude in order to maintain airspeed.

A high proficiency level occurs when you continually work toward a specific goal.

The CAAPVAA approach requires good thinking, planning, and airmanship. Just keep trying for as long as you fly airplanes, because you will be amazed at the results. When consistent performance occurs, your proficiency and mental awareness have reached an extremely high level.

CFI notes

The value of the CAAPVAA approach is impossible to ignore.

While there are innumerable ways to fly a landing approach, the value of the CAAPVAA approach is impossible to ignore. The mental aspects of this technique parallels the game of golf. Golfers have one grooved swing and a bag full of clubs. When students learn to fly the CAAPVAA approach, their mind becomes grooved and they can shoot par or better by using their *bag of tricks*—descent angles; ground track; flap extension; and knowledge of wind, updrafts, and downdrafts.

I stated previously that CAAPVAA complies with my training philosophy. Forced landings and night landings are the worst landing situations that a pilot will encounter, so you should teach CAAPVAA starting on day one. If you do this, your students will learn another important secret for establishing the proper level of self-confidence and flight safety.

Please remember that students will not master CAAPVAA until they are well past their first solo flight—it is not a presolo requirement. Until they learn to judge the factors involved, students will get high or low on approaches, and that mandates pitch and power corrections. CAAPVAA is something that good pilots work at for as long as they fly airplanes.

NORMAL LANDINGS

THE THREE STAGES OF A GOOD LANDING

Purpose

To simplify the teaching and understanding of landings.

1. From full flap extension to the flare point
2. From the flare point to touchdown
3. From touchdown until reaching taxi speed

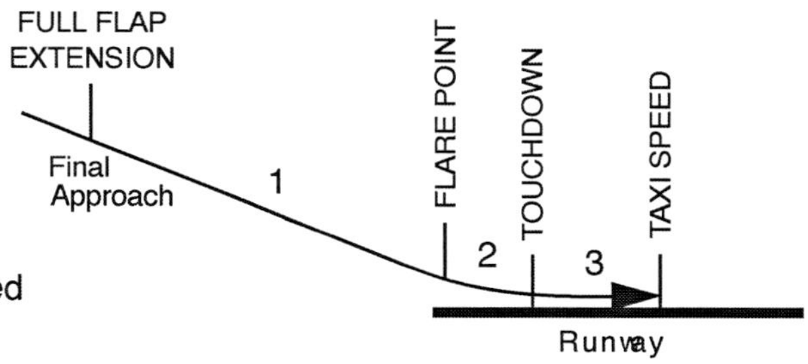

The Three Stages of a Good Landing

Stage 1—From Full Flap Extension To The Flare Point

Object

- Descent angle.
- Airspeed control.
- Runway centerline track.
- Flare point.

Execution

Full flaps are extended and you are confident that the current descent angle, attitude, and power setting will maintain airspeed and take you to the flare point. As the runway gets closer, your primary reference shifts from the descent angle to your planned touchdown point. The proper use of this reference requires you to note its relative position on the windshield and then keep it there. This is easy to do when you fly a constant attitude approach, but it becomes more difficult when you are forced to make pitch changes during this landing stage. If the touchdown point moves up on the windshield, you are too low; if it moves down, you are too high.

If the touchdown point moves up on the windshield, you're low; if it moves down, you're high.

Updrafts and downdrafts can negate your attempt to maintain constant attitude and power. If the altitude displacement becomes significant, use power to restore the proper descent angle, and simultaneously use pitch to maintain airspeed—elevator-throttle coordination. If you suddenly lose altitude, immediately add power to regain the lost altitude and increase pitch to maintain airspeed. If you suddenly gain altitude, immediately reduce power and decrease pitch to maintain airspeed. When you return to the proper descent angle, restore the previous attitude and power setting.

Runway centerline control means drift control. Unless you know the wind conditions for certain, you must rollout on final approach with the airplane parallel to and on the extended runway centerline. Now watch for drift. If it occurs, use bank angle to stop the drift and remain over the runway centerline; use rudder to remain parallel with the runway centerline. This is only way to evaluate actual wind conditions. Tower-reported winds are historical information.

Use bank angle to stay over the runway centerline; use rudder to stay parallel to the runway centerline.

Yes, if you are on an extended final approach, you should crab into the wind to control ground track and remain on the runway's extended centerline. But, as you get lower and start to concentrate on your touchdown point, you must transition to a sideslip.

Vision

- Airspeed and the touchdown spot.
- The sight picture.

Fixating your sight on any one reference will cause a bad landing.

During most of this landing stage, you continually cross-check airspeed and the touchdown spot in order to evaluate airspeed, descent angle, and drift. Fixating your sight on any one reference will cause a bad landing.

As you approach the runway and determine that you will definitely land close to your planned touchdown spot, look up at the focal point shown in the following illustration, but *maintain* the existing pitch attitude. You must master this discipline. Never let the airplane's pitch attitude increase when you look up at the focal point. If the nose rises one-quarter of an inch, you will lose five knots of airspeed, and that's unacceptable. The landing flare does not start until you lose more altitude.

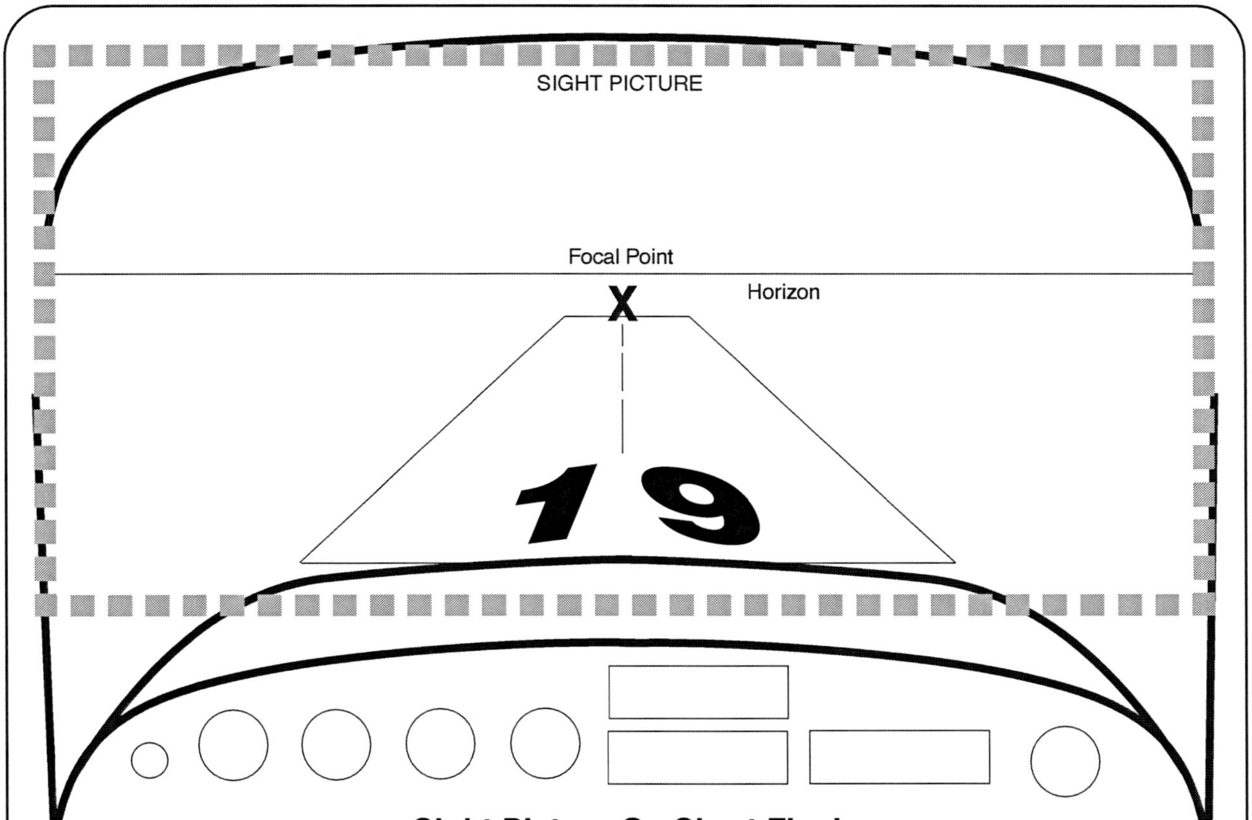

Sight Picture On Short Final

Now evaluate the sight picture, the total picture that your mind perceives when your eyes are pointed at the focal point. You see everything without fixating on any one thing.

Stage 2—From The Flare Point To Touchdown.

Object

- Recognize the flare point.

- Proper use of the forward and pie-in-the-sky sight pictures to evaluate sink rate, attitude, and drift.

- Recognize ground effect.

- The touchdown.

Execution

At the flare point, the runway starts to quickly widen. This is an important reference that works for all landings, day or night. An analogy of this occurs when you drive your car. You can see specific details when you look far ahead, but when you look over the hood, everything is blurred. Further ahead is the point where the blurring begins.

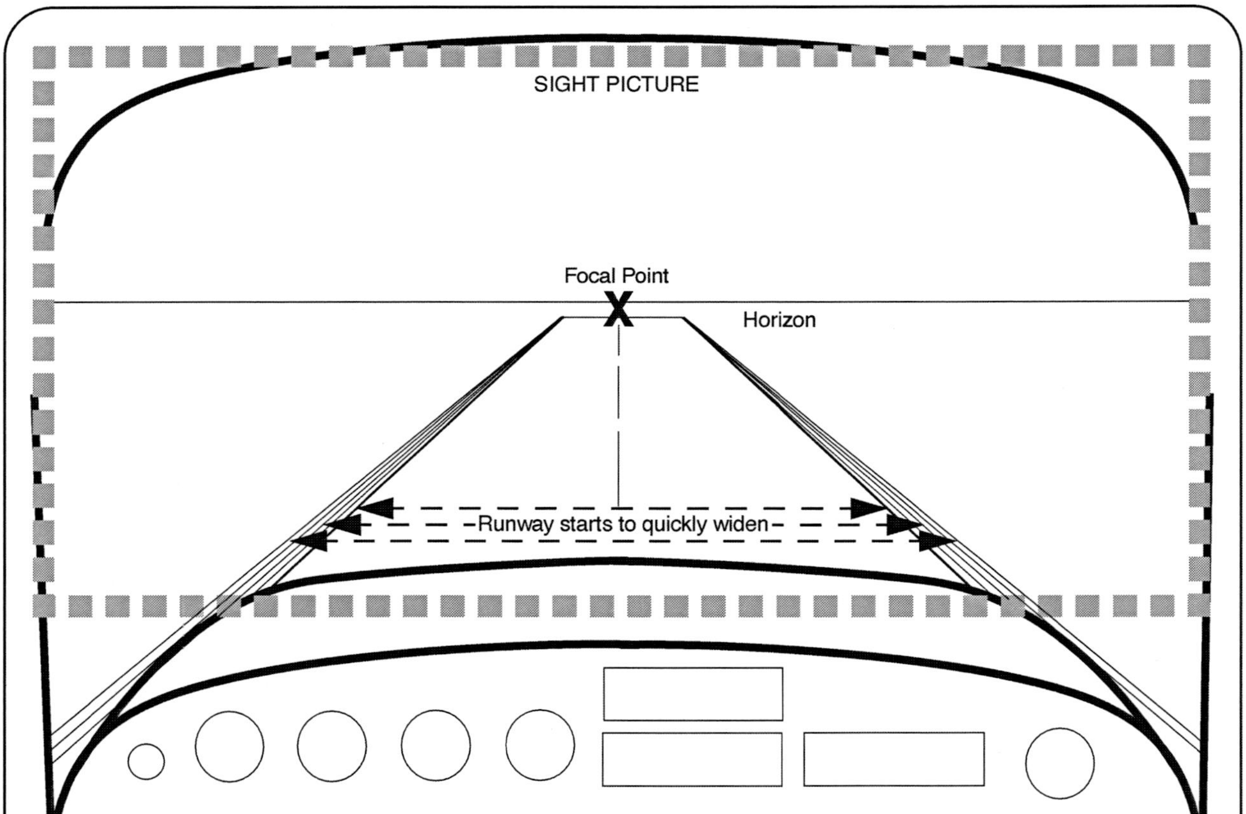

Sight Picture At The Flare Point

Another analogy occurs when you drive on a freeway and approach an underpass. The overhead structure just seems to sit in place when it is well ahead of you, but as you approach it, it quickly widens and accelerates toward you. This is what you want to observe as you approach a runway for landing. When the runway starts to quickly widen, you have reached the flare point.

At the flare point, your eyes are still pointed at the focal point, but your mind is evaluating the entire sight picture and paying particular attention to the area in the previous illustration that says, "runway starts to quickly widen." When that happens, start the landing flare, close the throttle, and then shift your eyes to the pie-in-the-sky sight picture that is shown in the following illustration. Now, as you slowly increase pitch to the landing attitude—which blocks your forward vision—you have sufficient ground references for proper evaluation.

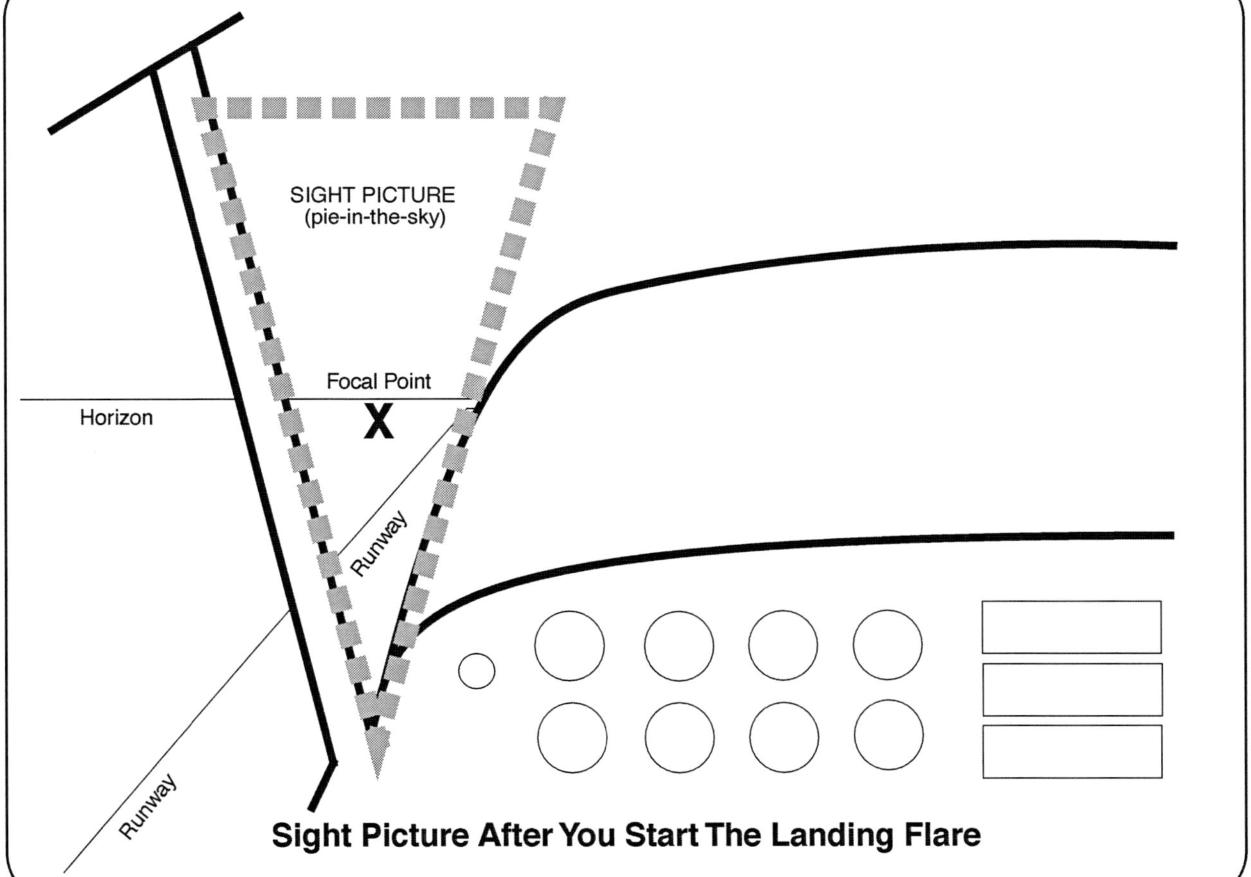

Sight Picture After You Start The Landing Flare

When touchdown is imminent, reverse your thinking and try to keep the airplane from landing. Hold it off the runway by slowly increasing pitch attitude. If done properly, the stall warning horn will activate just prior to touchdown.

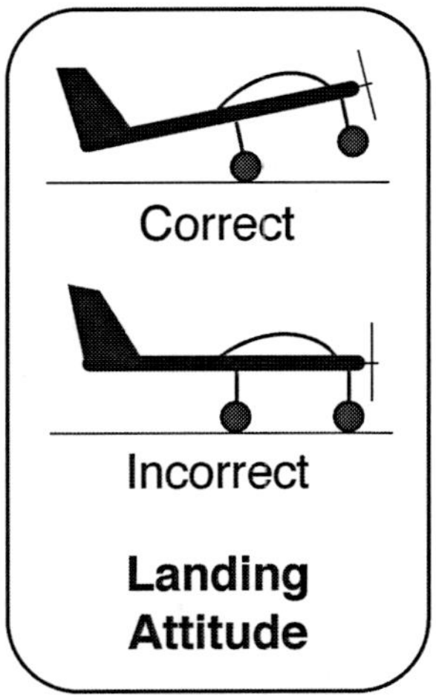

At touchdown, two critical actions must occur: The airplane *must be* in a nose-high attitude so that the nose wheel is well clear of the runway, and you must quickly move the yoke slightly aft in order to counteract the pitch-down that occurs when the main wheels contact the runway. The nose wheel will settle onto the runway automatically as airspeed decreases, and the aerodynamic drag from the nose-high attitude will help you slow down without using the brakes.

If you increase pitch attitude too fast during the flare, you will *balloon*. This is a quick increase in altitude which is extremely dangerous when you are at low speed close to the ground. Go-around if you balloon excessively. Add some power and re-establish the landing attitude if the deviation is minor.

Wheelbarrowing is a major cause of landing accidents.

If you increase pitch attitude too slow, a firm landing will occur, or in the worst case, you will wheelbarrow. This occurs when an airplane lands on the nose wheel, and it is a major cause of landing accidents. A tricycle-gear airplane's center of gravity is located just forward of the main wheels. If the airplane lands on its nose wheel, that wheel becomes a pivot point for the center of gravity, which can generate a violent, unrecoverable yaw and an embarrassing accident.

Vision

Shift your eyes to the pie-in-the-sky sight picture just before the nose cowl blocks your forward sight picture.

Do not wait until the nose cowl blocks your forward sight picture and then shift your eyes to the pie-in-the-sky sight picture. You will lose your visual perceptions, and it will take a moment or two for you to re-establish them. That interruption will spoil what would have been a perfect landing. You must shift to the pie-in-the-sky sight picture just before you lose your forward sight picture and those visual perceptions.

The focal point should move closer to the airplane as speed decreases. Compare the focal-point locations in relation to the horizon in the last two sight-picture illustrations. The focal point in the last illustration will move down toward the edge of the runway just prior to touchdown.

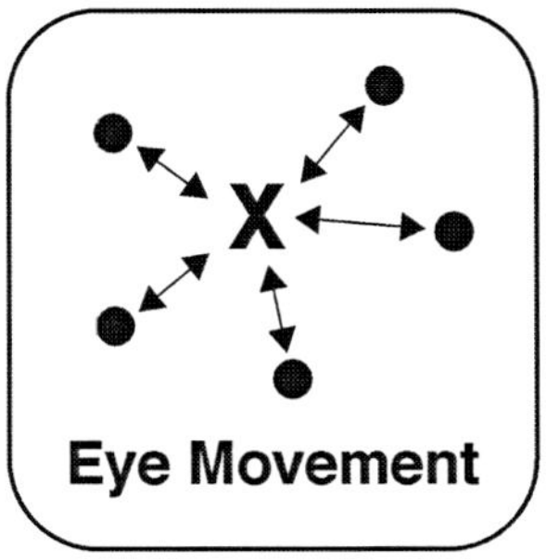

Slight, continuous eye movement increases visual acuity. The illustration on the left gives an example. The **X** is the focal point, but your eyes should dance back and forth at random to the various dots. No specific order is required. Just keep your eyes moving in the vicinity of the focal point.

Here are a few additional vision rules:

- Eye orientation when flying is completely different from what you do on the ground. You must develop these new habits in order to fly properly.

- Never lose sight of the Earth's surface.

- Use the proper sight picture.

- Looking too close or too far from the airplane will cause inconsistent landings.

- Pitch attitude will decrease when an airplane enters ground effect. Don't let that happen. If you maintain attitude, speed will increase due to the decreased drag that occurs in ground effect and sink rate will decrease, which aids the landing process.

- As touchdown becomes imminent, the sink rate should be negligible, and the airplane must be in the landing attitude. if you do this, your touchdown will be extremely smooth—that's called a "greaser."

- If you do not shift your eyes to the pie-in-the-sky sight picture as instructed, the following will occur:

Good landings will be the result of luck rather than skill.

You will not make consistently good landings or have the proper self-confidence when you transition to an unfamiliar airplane.

You will develop an elongated neck from trying to look over the airplane's nose cowl. Caution! If this happens, you may wind up on display in the giraffe habitat at your local zoo.

Stage 3—From Touchdown Until You Reach Taxi Speed.

Object

• Keep the airplane's weight on the main wheels during deceleration in order to prevent wheelbarrowing.

• Stay on the runway centerline until you reach taxi speed and the runway's exit point.

Never relax control input until taxi speed has been reached.

The airplane landed and you immediately moved the yoke back slightly in order to keep the nose from pitching down. The nose wheel will automatically settle to the runway when airspeed and elevator effectiveness decrease. Maintain directional control with rudder, and position the ailerons for the existing wind condition. Never relax control input until taxi speed has been reached.

Keep the yoke back—elevator up—and stay off the wheel brakes when you are moving at high speed on the ground. This maximizes the weight that is on the main wheels and minimizes the weight that is on the nose wheel. You will skid the tires if you apply brakes immediately after touchdown, because the wings still support most of the airplane's weight.

Vision

• The pie-in-the-sky sight picture until nose wheel touchdown.

• Straight ahead after nose-wheel touchdown. Do not look at or do anything inside the cockpit until you taxi clear of the runway's hold-short lines.

CFI Notes

The CAAPVAA approach and the three stages of a good landing get excellent results. The old saying, "A good landing requires a good approach," does not register with students. They are obsessed with the landing, not the approach. So designate a simulated runway in the practice area or use the the runway at your airport, and practice flying from the downwind leg to short final using the CAAPVAA approach. Approaching the runway, execute a go-around. Do not allow landings until students can demonstrate that they understand how the approach and go-around should be executed.

Never solo a student who has a tendency to move the yoke forward and force the airplane onto the runway.

During landing training make certain that students understand the consequences of wheelbarrowing. Watch carefully for any tendency to move the yoke forward and force the airplane onto the runway or to move the yoke forward when traveling at high speed on the runway. Never solo a student who has not totally eliminated this dangerous error.

If students do not shift their eyes to the pie-in-the-sky sight picture during the landing flare, quickly cover the windshield with a manila file folder or navigation chart in order to enforce this important prerequisite for consistent, safe landings.

Landing at busy airports requires teamwork between all pilots and controllers. When clear of the runway, teach your students to call for taxi clearance and then perform the after-landing, clean-up actions. This keeps airport traffic moving.

GO-AROUNDS

Purpose

To abort a landing due to a bad approach, traffic separation, or a safety consideration such as wake turbulence.

You or the air traffic controller initiate the go-around. This is a normal flight maneuver that reflects good judgment. Never attempt to salvage a bad approach or landing. Go-around!

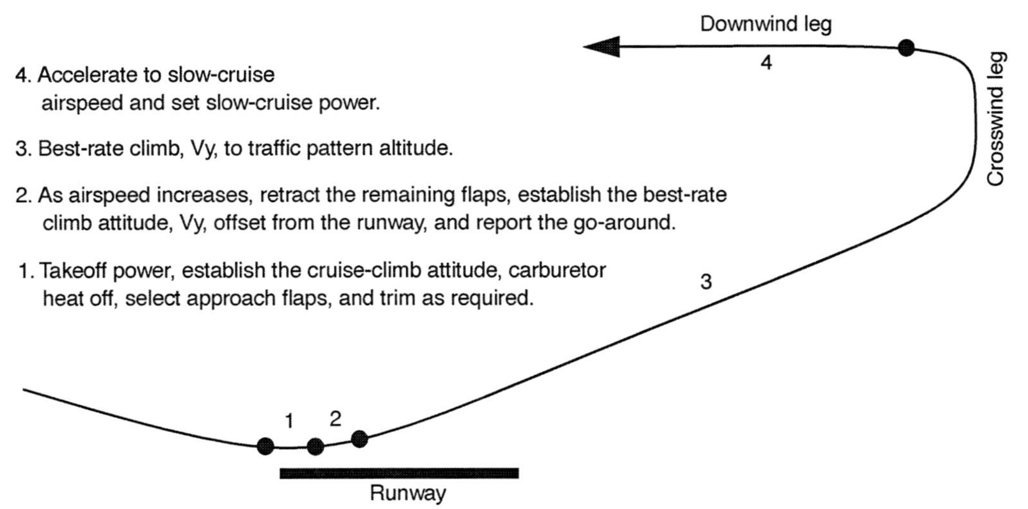

The Go-Around

Execution

Clear the surrounding airspace for traffic. When you decide to go-around, you simultaneously establish the cruise-climb attitude and set full power, turn the carburetor heat off, select approach flaps, and rough-trim the elevator. When airspeed approaches V_Y, establish the best-rate climb attitude, and retract the remaining flaps.

Never talk on the radio until after you retract the flaps and establish the climb.

If an airplane on the runway caused the go-around, offset your ground track from the runway—preferably to the right side—in order to keep that airplane in sight. Finally, report the go-around, and continue the best-rate climb to pattern altitude. Never, absolutely never, report the go-around on the radio until after you establish the climb and retract the flaps.

Attitude

Immediately establish the T-pitch attitude for initial acceleration. Because the elevator was previously trimmed for the approach attitude, it will be grossly out of trim for the go-around. As you apply full power, you must aggressively establish the T-pitch attitude, regardless of the elevator control force. As you approach V_Y, establish the I-pitch attitude for best-rate climb and retract the remaining flaps.

Power

Full power.

Trim

Rough-trim as necessary. Fine-trim after establishing the best-rate climb with flaps retracted.

Vision

Three points must be scanned. Evaluate these references while you scan the surrounding airspace for traffic:

- Use the wingtip and the horizon to establish pitch attitude and maintain bank attitude.

- Use the side of the nose and the horizon to establish bank attitude and maintain pitch attitude.

- Use the pie-in-the-sky sight picture to supplement the first two references and evaluate roll, pitch, and yaw.

CFI Notes

Students must understand that a go-around is a normal maneuver that demonstrates good judgment and awareness.

Introduce go-arounds at altitude, and point out that this maneuver parallels the recovery procedure for slow flight and power-off stalls. Then have students perform the following:

• Select approach flaps before they add power and increase pitch for the go-around. Note the immediate sink.

• Retract all of the flaps when they add power for the go-around. Note the immediate sink rate and the possible activation of the stall warning horn.

• Set full nose-up elevator trim and perform a go-around. Note the rapid rate of pitch increase, the airplane's tendency to establish an excessive nose-high attitude, and the high elevator force that is required to maintain the T-pitch attitude.

• Make a go-around without using rudder, and note the left turning tendency of the airplane.

• Establish a right sideslip in the landing configuration—to simulate a right crosswind—and execute a go-around. Note the rapid left turning tendency of the airplane, because the rudder is displaced to the left.

RECOVERY FROM BAD APPROACHES AND LANDINGS

Purpose

To recognize and correct improper approaches and landings. To relate airspeed to control feel, sink rate to power, and visual perception to the sight pictures.

Execution

Low or Slow Final Approach

• If low and control pressures feel mushy, add power, maintain the C-pitch attitude until control pressures feel normal, and then increase the pitch attitude to regain altitude.

• If low and control pressures feel normal, simultaneously add power and increase the pitch attitude to regain altitude.

- If low and control pressures feel firm, increase the pitch attitude to regain altitude.

High Sink Rate

- If control pressures feel mushy, make a large power increase and increase pitch attitude. When sink rate is normal, reestablish the C-pitch attitude, but do not decrease power until control pressures feel normal.

- If control pressures feel normal, make a moderate power increase and increase the pitch attitude. When sink rate is normal, reestablish the C-pitch attitude and decrease power.

- If control pressures feel firm, increase the pitch attitude.

High Level-off

- If control pressures feel mushy, establish the C-pitch attitude and add power to prevent further airspeed decrease. Descend to the proper height, and resume the flare.

- If control pressures feel normal, establish the C-pitch attitude, descend to the proper height, and resume the flare.

Late or Rapid Level-Off

- If control pressures feel mushy, add power. Slowly reduce power, establish the landing attitude, and let the airplane settle onto the runway.

- If control pressures feel normal, establish the landing attitude, and let the airplane settle onto the runway.

Floating During Landing

If you float excessively, go-around.

- Close the throttle, and let the airplane settle onto the runway as you establish the landing attitude. If you float excessively, go-around.

Hard Landing

- Close the throttle, maintain the landing attitude with elevator, and maintain directional control with rudder.

- *Never move the yoke forward!* Hold the yoke back to keep the airplane's weight on the main wheels. Never relax the elevator back pressure until you have decelerated to taxi speed.

Shallow Skip Landing

- If control pressures feel mushy, maintain the existing pitch attitude and add enough power to prevent further airspeed decay. Let the airplane settle onto the runway as you establish the landing attitude and reduce power.

- If control pressures feel normal, let the airplane settle onto the runway as you establish the landing attitude.

High Bounce Landing

Go-around whenever a landing deviation is excesive.

- Go-around. However, if runway length is sufficient, decrease pitch attitude to level off and add considerable power. Reduce power to resume the descent to the runway, and let the airplane settle onto the runway as you establish the landing attitude.

Crabbed or Drifting Touchdown

- Quickly move the yoke aft to keep the nose wheel clear of the runway.

- Remove crab with rudder. Stop drift with the ailerons. If you are landing in a crosswind, increase the aileron displacement at touchdown in order to maximize the down force on the upwind main wheel.

TOUCH-AND-GO LANDINGS

Purpose

Touch-and-go landings violate three standard operating rules.

This maneuver has many benefits. It forces students to display proper multitasking skills, attitude control, control coordination, awareness, and judgment; but it does violate three standard operating rules.

Execution

After landing, maintain elevator back pressure, retract the flaps, reset the elevator trim, turn off the carburetor heat if applicable, and set full power.

Hazards

- Touch-and-go landings can lead to wheelbarrowing and loss of control if elevator back pressure is relaxed while rolling down the runway. Maintain elevator back pressure while you are moving at high speed on the runway.

- In the rush to configure the airplane for takeoff, students have forgotten to retract the flaps, reposition the elevator trim, or turn off the carburetor heat, and they have inadvertently retracted the landing gear instead of the flaps in airplanes that have retractable landing gear.

- If a problem or judgment error occurs after full power has been set, an aborted takeoff within the confines of the runway may be impossible.

Violated Operating Rules

- *Add power before retracting the flaps.* This rule is used during go-arounds and during the recovery from slow flight and flaps extended stalls. Touch-and-go landings violate this rule because the flaps are retracted before power is added.

- *Never look in the cockpit while moving at high speed on the runway, and for positive identification, look at the switch or control you are about to activate.* These rules are violated, because during touch-and-go landings, students usually glance at the flap and carburetor heat selectors. The elevator trim selector can be identified by feel, as can the flap and carburetor heat selectors in most cases.

- *Never activate cockpit controls on the runway.* Touch-and-go landings obviously violate this rule. If you turn the carburetor heat off on short final, you eliminate one distraction.

FLYING SOLO

Your first solo! A moment never forgotten. It occurs when your instructor recognizes consistent performance combined with good judgment and awareness.

Your first three solo flights are supervised by your instructor. On the first flight, your instructor will get out of the airplane, and watch you make three landings—two touch-and-go and one full stop. On the second flight, you make touch-and-go landings for the remainder of the period after your instructor gets out of the airplane. The third flight is entirely solo, but your instructor will watch your performance from a convenient location on the airport.

The quickest way to lose solo flying privileges, is to ignore your instructor's instructions or show poor judgment.

After the three supervised solo flights, your instructor will periodically authorize you to make additional solo flights. The quickest way to lose solo flying privileges is to fly without your instructor's specific approval, ignore your instructor's instructions, or show poor judgment.

Solo flight gives you a chance to practice maneuvers on your own and build self-confidence. You can also develop bad habits or improper techniques during these flights, and for that reason, solo flying should be kept to a minimum.

REVIEW TEST

1. What is the purpose of the departure profile, and how is it flown?
2. What is the safe altitude?
3. Describe the rudder and elevator inputs that are required during the takeoff roll.
4. If you don't move the yoke forward after liftoff, the airplane will pitch up. Why?
5. Why must full throttle be used for climb power when specified by the POH?
6. If you plan to remain in the traffic pattern after takeoff, when do you start your turn onto the crosswind leg?
7. Describe what you must do at the traffic pattern's view point, check point, and key point.
8. If you are following other traffic in the pattern, when do you start your descent?
9. If told that you are number four to land, what is the quickest way to locate your traffic?
10. Why should you use the CAAPVAA?
11. What tools are used to fly the CAAPVAA?
12. If high on base leg, how should you use bank attitude to correct?
13. If high on final approach with full flaps extended, how should you correct?
14. If high with all corrections applied and you think you might be able to land and stop before the end of the runway, you should?
15. How can you anticipate vertical air movement and wind gradient changes?
16. List the three stages of a good landing.
17. What are you visually evaluating during each landing stage?
18. What will occur if you push the yoke forward and try to force the airplane onto the runway?
19. Approaching the runway, how do you manage the aileron and rudder controls?

20. When can you turn off the runway centerline?
21. List the sequence of events for executing a go-around.
22. Define a go-around and state your last priority when you execute this maneuver.
23. During the landing rollout, where should you be looking?
24. If low or slow on final approach and control pressures feel mushy, you should?
25. If experiencing a high sink rate and control pressures feel mushy, you should?
26. If experiencing a high level-off and control pressures feel mushy, you should?
27. If experiencing a late level-off and control pressures feel mushy, you should?
28. If a hard landing occurs, you should?
29. If a shallow skip landing occurs and control pressures feel mushy, you should?
30. If a high bounce landing occurs, you should?
31. If a crabbed touchdown occurs, you should?
32. What are the hazards of touch-and-go landings?
33. What standard operating procedures are violated during touch-and-go landings?
34. When flying solo, what is your primary obligation?

IF A PASSENGER IS NOT HAPPY WITH
A LANDING, IT'S NOT THE PILOT'S FAULT!
IT WAS DUE TO ONE OF THE FOLLOWING
ELEMENTS THAT PILOTS CAN'T CONTROL:
RUNWAY SKEW,
RUNWAY REPULSION,
OR RUNWAY ATTRACTION.

CHAPTER 5

ADVANCED MANEUVERS FOR PRIVATE PILOT CERTIFICATION

AIRPLANE INSTRUMENTATION

Instruments are not needed to fly an airplane safely in visual conditions (clear of clouds).

Never forget this fact: To fly an airplane safely in visual conditions, no instruments are needed. The wing and throttle positions (attitude and power), engine and air noise, and control feel provide the basic information for safe flight. Once you develop those insights, you can add instrument reference.

What is the most important cockpit instrument? When I ask pilots that question, they usually ignore common sense and quote federal regulations. To answer that question properly, assume that I gave you a brand new airplane that had no instrumentation whatsoever. What instruments would you add and in what sequence? That will answer my initial question. Here are my instrument priorities, listed in their relative order of importance:

- The magnetic compass for orientation and navigation. It is the most important cockpit instrument, because without it, most flights are confined to the local area.

Fuel gauges are never trustworthy.

- The clock for navigation computations and fuel consumption. Fuel gauges are never trustworthy.

- The oil pressure and oil temperature gauges for engine health.

- The tachometer for engine performance and limitations.

- The altimeter for traffic separation and altitude. Without an altimeter, you cannot fly more than 3,000 feet above ground level (agl) due to altitude rules for traffic separation.

- The airspeed indicator for airspeed and limitations.

- The remaining required instruments.

INITIAL AIRWORK

TURNS TO MAGNETIC COMPASS HEADINGS

This maneuver is not required for private pilot certification, but all pilots should understand magnetic compass characteristics and operating rules.

Purpose

Remember, the magnetic compass is your most important instrument.

To master the characteristics and operating rules for this faithful instrument, the airplane's master heading reference. To remain oriented and fly a specific magnetic heading when navigating a designated course.

Characteristics

The magnetic compass is filled with non-acidic, white kerosene that dampens card movement, reduces friction, and prevents freezing. Kerosene must be added whenever the fluid level is visible in the compass window.

Mechanics periodically "swing" the compass to ensure accuracy. To do this, they position the airplane over a compass rose—a reference circle located on the airport that is marked with known magnetic headings—and adjust the compensating magnets—located in the compass case—so that heading errors are minimized when the engine and all avionics equipment are operating. The errors that remain are listed on the compass deviation card that is attached on or near the compass housing.

Never trust a magnetic compass until its accuracy has been verified.

Never trust a magnetic compass until you have verified its accuracy by taxiing parallel to and perpendicular to a runway. When you use the compass, the magnetic field in the cockpit must duplicate that which existed when the compass was swung. Therefore, you must keep metal objects well clear of the compass and turn on all avionics equipment.

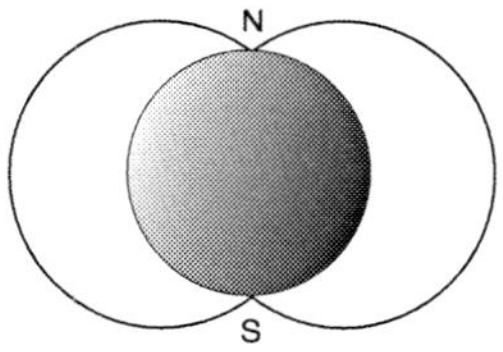

Earth's Magnetic Lines of Force

The errors exhibited by the compass card in flight are caused by magnetic-dip, an action that causes the north-seeking end of the compass needle—and the compass card to which it is attached—to tilt down as the airplane moves away from the Earth's equator. This occurs because the compass needle always remains parallel to the Earth's magnetic lines of force. These lines are parallel to the Earth's surface at the equator, but form progressively steeper angles as they move away from the equator. Near the Earth's poles, the angle becomes so steep that the magnetic compass is useless.

When the compass needle (and card) tilt due to acceleration, deceleration, banking, or turbulence, the north-seeking end of the needle moves toward the Earth's surface, which turns the card and generates the following two errors when you fly in the Earth's northern hemisphere. (These errors are reversed in the southern hemisphere.)

Acceleration and deceleration errors They are non-existent when you fly on a north or south heading, but they are maximum when you fly on a west or east heading. The mnemonic *ANDS* is used to remember the following relationship: **A**cceleration, the card turns toward **N**orth; **D**eceleration, the card turns toward **S**outh. These temporary errors disappear when you stop accelerating or decelerating.

The northerly turning error This occurs when you bank the airplane. If you are turning to a heading that is located on the northern half of the compass card, the card lags behind the airplane's actual heading; if turning to a heading that is located on the southern half of the compass card, the card leads the airplane's actual heading. Four rules allow you to compensate for this error.

- The banking rule: Do not exceed the A-bank attitude. 18 degrees is the bank limit for most magnetic compasses.

To perform magnetic compass turns to specific headings, memorize the banking, latitude, turning, and rollout rules.

• The latitude rule: The Earth's latitude at the airplane's position represents the maximum turning error. An airplane flying over Los Angeles, California, 33-degrees north latitude, experiences a maximum error of 33 degrees. An airplane flying over Seattle, Washington, 47-degrees north latitude, experiences a maximum error of 47 degrees.

• The turning rule: When turning to due north or south, the latitude-rule error is maximum; to due east or west, the error is zero; and to an intermediate heading, the error is proportional as is depicted in the following illustration for an airplane located at 30-degrees north latitude.

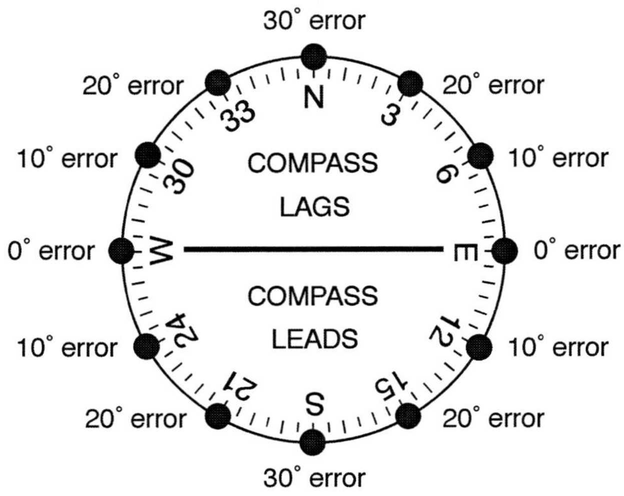

Magnetic Compass Turning Errors At 30-Degrees North Latitude

• The rollout rule: If you are turning to a heading that is located on the northern half of the compass card, rollout before you reach it. If the heading is located on the southern half of the compass card, rollout after you pass it.

To determine the rollout point, apply the turning error to the assigned heading.

To determine the rollout point, apply the turning error to the assigned heading. The direction of turn makes no difference. For example, if turning to 030 degrees in the previous illustration, stop your turn 20-degrees short of 030 degrees—either 010 or 050 degrees. If turning to an assigned heading of 180 degrees, stop your turn 30-degrees past 180 degrees—either 150 or 210 degrees.

To remember the rule, use this relationship: Warm air is good, go south—rollout after you pass the new heading if it is located on the southern half of the compass card. Cold air is bad, don't go north—rollout before you reach the new heading if it is located on the northern half of the compass card.

BEST-ANGLE CLIMB (V_X)

Purpose

To gain the most altitude with the least forward travel. To climb as steeply as possible. Use this climb when attempting to clear obstacles at the end of a runway during takeoff or clear terrain or clouds on cross-country flights. The following illustration compares the three standard climb procedures.

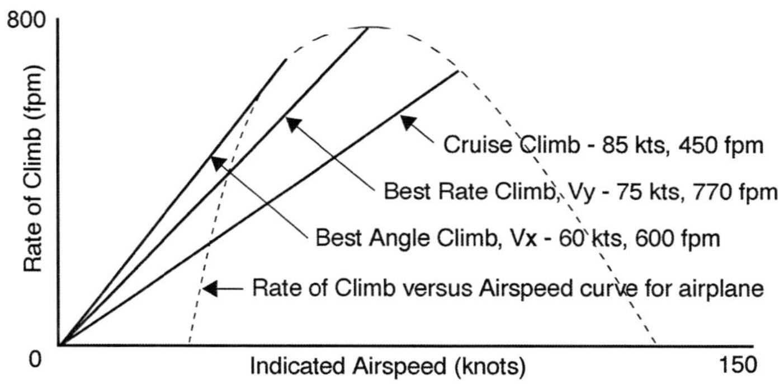

Comparing Vx, Vy, and Cruise Climbs

Attitude

Establish and maintain the P-pitch attitude and the B-bank attitude. If your present airspeed is at or above best angle of climb airspeed, select the pitch attitude immediately. To terminate the climb, you establish and maintain the desired climb or cruise attitude.

Power

Full power. If your present airspeed is above the best angle of climb airspeed, do not increase power until airspeed decreases to that speed. When terminating the climb, let airspeed accelerate to the desired cruise speed before you reduce power.

Trim

Rough-trim the elevator whenever elevator pressures get heavy. Fine-trim the flight controls, when airspeed and power have stabilized.

Vision

Three points must be scanned. Evaluate these references while you scan the surrounding airspace for traffic:

- Use the wingtip and the horizon to establish pitch attitude and maintain bank attitude.

- Use the side of the nose and the horizon to establish bank attitude and maintain pitch attitude.

- Use the pie-in-the-sky sight picture to supplement the first two references and evaluate roll, pitch, and yaw.

STEEP TURNS

Purpose

To perform right and left 360-degree, steep-bank turns while you maintain altitude and orientation.

Execution

Pick a reference spot on the horizon (do not use the heading indicator) and start the turn. To maintain altitude, you must increase elevator back pressure as bank angle increases and decrease it as bank angle decreases. End the maneuver at the original reference spot.

A mechanical correction is one that you make before the actual need for that correction becomes apparent.

During the entry, you must make two mechanical corrections—a correction that you make before the need for it becomes apparent. As you roll through 30 degrees of bank, make an audible power increase. As you reach 45 degrees of bank, immediately increase elevator back pressure until you feel 1.4-Gs. Remove those two corrections when you start your rollout so that altitude and airspeed do not increase at the completion of the maneuver.

The power increase keeps airspeed from decreasing significantly (stall speed increases as bank angle and load factor increase), and it pitches the nose up, two factors that make the maneuver easier to fly. Never look at the tachometer for this correction. Just note the amount of throttle movement, because you must remove it when you start the rollout.

You must be aware of the G-load in order to fly the maneuver properly. During a level, 45-degree bank turn, 1.4-Gs occurs in all airplanes. During a level, 60-degree bank turn, 2-Gs occur. The stall speed increase is the square root of the load factor (G-load).

Airplanes do not react instantaneously.

Use the elevator to maintain altitude and G-load during the turn. For very minor *vertical speed* deviations—yes, for maintaining altitude the vertical speed indicator is more useful than the altimeter—you can momentarily steepen or shallow the bank no more than five degrees. Make small corrections and wait a moment to evaluate the response. Airplanes do not react instantaneously.

For example, vertical speed indicates a slight climb. Steepen the bank angle slightly to stop that trend, and then restore the 45-degree bank with a little less elevator back pressure (G-load). Or, if vertical speed indicates a slight descent, shallow the bank angle slightly to stop that trend, and then restore the 45-degree bank with a little more back pressure.

Caution! Keep your eyes on the visual references outside the cockpit during this maneuver. You will, however, occasionally glance at the vertical speed indicator.

Attitude
The C-pitch attitude and the K-bank attitude.

Power
Cruise or slow cruise power that is temporarily increased during the maneuver.

Trim
Do not retrim the airplane during the turn.

Do not retrim the elevator during the turn, because that adds additional drag and creates an unnecessary distraction. The power increase offsets the turn's drag increase and eliminates excessively high elevator pressures that would make an elevator trim change seem appropriate.

Vision

Clear the surrounding airspace for traffic. Before you start the turn, look straight ahead at your reference point. That line of sight must be parallel to the airplane's longitudinal axis, not angled toward the airplane's nose spinner. Note the spot where that line of sight passes through the windshield. Hopefully, there is a bug or dirt spot on the windshield at that location. If not, your instructor can make a spot using a grease pencil or a small piece of tape. While you roll into and out of the turn, you must watch the reference spot and not let it rise or descend in reference to the horizon.

During the turn, the horizon is usually uneven due to hills and mountains, which makes it difficult to evaluate vertical movement of the windshield reference spot. To compensate, you must monitor the movement of the windshield's framework in relation to the Earth's surface. Yes, you are looking at the windshield reference spot and the horizon, but you are also monitoring the windshield's framework and its movement relative to the surface. You must use a sight picture just like you were taught to do when landing.

CFI Notes

Because this is not a constant airspeed maneuver, some instructors feel that a power increase is not required. Students are told to use elevator trim. Yes, that can be done, but such thinking is incorrect.

Adding power during steep turns enhances safety.

If power is not added, airspeed decays during the first half of the turn due to the drag increase that is generated by the increased lift and the increased elevator and trim tab displacements. Airspeed is decreasing and stall speed is increasing as the bank angle increases. The power increase offsets that drag increase, reduces the required elevator force, and enhances safety for low-time pilots.

Students will look in the direction of turn, clear the airspace, and start the steep-bank turn while still looking to the side. Never allow that to occur. They must be looking straight ahead during both the entry and the recovery. If their line of sight angles toward the prop spinner, they will climb during right turns and descend during left turns.

STALLS AND SPINS

CONTINUOUS STALL EXERCISE

Purpose
A coordination exercise that demonstrates rudder effectiveness and aileron ineffectiveness during low speed flight.

Execution
Step on the high or rising wing with rudder input.

Clear the surrounding airspace for traffic. Enter a power-off, flaps-up stall, but keep the elevator fully deflected and the ailerons neutral until the exercise is terminated. Most modern training airplanes will porpoise up and down and attempt to roll to the right or the left. Use rudder to prevent the roll and keep the airplane flying straight. When mastered, make shallow left- and right-banked turns using only the rudder. Remember this rule: Step on the high or rising wing—right wing, right rudder; left wing, left rudder.

You must apply the two-step correction rule to this maneuver. When the airplane starts to roll, stop the roll with opposite rudder (full rudder may be required). At the start of a roll in the opposite direction, immediately remove half of that rudder input and bring the airplane back to a wings-level attitude.

ACCELERATED STALLS

Purpose
To prove that stalls can occur at any airspeed. To experience the airplane's accelerated stall and recovery characteristics. To develop attitude awareness.

Execution
Establish a level 45-degree bank turn and set power so that airspeed is well below the airplane's maneuvering speed but at least 20 knots above the unaccelerated stall speed. Quickly increase elevator pressure until the stall occurs. To recover, immediately release the elevator back pressure and add power if needed.

Attitude
The K-bank and T-pitch attitudes. Recover by simultaneously establishing the B-bank and C-pitch attitudes.

Power
Reduced to establish the entry airspeed. Full power if needed for the recovery.

Trim
As required when the entry airspeed is established.

Vision
Clear the surrounding airspace for traffic. Use the wing and nose positions to set and maintain attitude. Use the Earth's surface and the horizon, framed by the nose cowl and the windshield's framework, to evaluate roll, pitch, and yaw.

CROSSED-CONTROL STALLS

Purpose
To demonstrate airplane reaction when stalled with crossed controls—opposite rudder and aileron. To understand the two situations that lead to crossed-control stalls.

Execution
Clear the surrounding airspace for traffic. Set up the two following landing-approach situations at altitude, and allow the airplane to stall.

The skidding turn to final This can occur when the airplane is low and turning from base leg to final approach. Due to the low altitude, the pilot, not wanting to bank excessively, attempts to turn with rudder. The excessive rudder input causes bank to increase and the pitch to decrease. The pilot compensates by holding the low wing and the nose up using increased aileron and elevator displacement. These control movements can result in a stall followed by a spin entry toward the inside of the turn.

To avoid these stalls, you must keep the control inputs coordinated.

The slipping turn to final This can occur when the airplane is overshooting the runway centerline while turning from base leg to final approach. The pilot rolls rapidly into a steep bank, but does not use enough rudder pressure. The steep bank causes pitch to decrease and sink rate to increase. The pilot compensates by pulling the yoke back. This control movement can result in an accelerated stall followed by a spin entry toward the outside of the turn.

CFI Notes

Demonstrate cross-controlled stalls at altitude with the flaps retracted. If flaps are extended for these demonstrations, you could exceed airframe limitations.

SPINS

Purpose

To develop confidence. To understand the effectiveness of rudder input during slow speed flight. To understand how spins occur, the spin entry, and the spin recovery. To develop a high degree of awareness for control coordination, aerodynamic responses, attitude reference points, and orientation.

Execution

Before flying, check the airplane's weight and balance data. Defuel if necessary and remove all loose items located in the cabin. A forward center of gravity improves spin performance.

Climb to 3,000 feet above ground level, clear the surrounding airspace for traffic, and bring the airplane into a relatively steep, nose-high attitude so that a crisp stall occurs. At the stall, close the throttle (if it was used for a power-on stall entry), apply full up elevator and full left or right rudder. During the spin, keep the elevator and rudder fully displaced, the throttle idled, and the ailerons neutralized.

To recover, move the controls quickly and forcefully so that positive recovery will be initiated. The following basic recovery procedure is not to be used in lieu of the recovery technique specified in the airplane's POH.

- Keep the throttle closed.
- Keep the ailerons neutralized.
- Apply and hold full opposite rudder.
- When spin rotation stops, quickly neutralize the rudder and move the yoke forward to break the stall.
- Recover from the ensuing dive without incurring a secondary stall or allowing an excessive airspeed increase.

On page 3 - 29, I stated that rudder input is the most effective way to control an unsolicited roll—a roll that is caused by external forces, not you. I cannot overstress the importance of making rudder input an automatic reaction to the roll rather than aileron displacement, which is contrary to your earthbound habit patterns. Rudder, not aileron, *must be* your first input when an unsolicited roll occurs during stalls or slow flight or during an encounter with wake turbulence or wind shear.

CFI Notes

Some modern, spinable airplanes are difficult to spin because of their inherent stability. To ensure a good spin entry, increase the pitch attitude at a fast rate so that a steep nose high attitude exists when the stall occurs.

Two other actions can be used to induce a spin: First, use a power-on stall entry and select idle power when the airplane starts rolling into the spin. This is known as a power-entry spin. Second, apply opposite aileron when the stall occurs but neutralize aileron input when the power is reduced to idle. This action helps the inside wing stall and start the spin rotation.

Spins may also be entered from medium banked turns. Using bottom rudder results in a quick, into-the-turn spin entry. Using top rudder results in a slow over-the-top entry.

If the student does not hold the yoke fully aft and keep the rudder pedal fully depressed, the spin will turn into a spiral with a rapid increase in airspeed. If this occurs, recover immediately. Remember, an airplane cannot spin if it's not stalled.

Do not force students to do spins. If they are apprehensive about flying, do not even suggest spins, but spend considerable time teaching stalls, the continuous stall exercise, and slow flight. To intentional scare students creates apprehension about airplane safety and reduces self-confidence. With proper training, apprehensive students can become safe pilots.

On the other hand, never refuse a student who wants to spin. Fifteen minutes of spins proves the value of our stall recovery procedures and greatly improves pilot self-confidence and awareness. Remember, spins are one of the easiest maneuvers to perform.

Spins are one of the easiest maneuvers to perform.

GROUND REFERENCE MANEUVERS

TURNS AROUND A POINT

Purpose

To circle a point on the ground at slow cruise while maintaining altitude and a constant turn radius. To recognize wind drift while turning and compensate by adjusting the bank angle. To plan and control a circular ground track while properly dividing your attention.

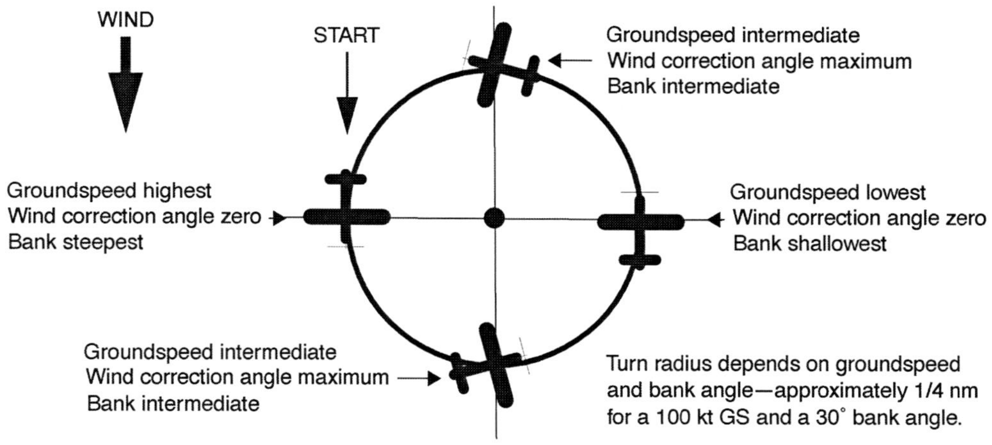

Turns Around a Point

Execution

Clear the surrounding airspace for traffic. Start the maneuver upwind of a well defined pylon at 800 to 1,000 feet agl. Fly downwind until the pylon is abeam either wing tip and start a medium bank turn. If a shallow bank is required, you are too far from the pylon; if a steep bank is required, you are too close. As you circle, adjust bank angle in order to fly over the desired ground track. Bank angle must vary in direct proportion to the constantly changing groundspeed.

Attitude

Pitch attitude and airspeed remain constant throughout the maneuver. Bank attitude, the wind correction angle, and groundspeed continually vary. Four reference points are used to illustrate these variations.

- At the starting point with a direct tailwind: Groundspeed and bank angle are greatest. Start with the N-bank attitude. The airplane is flying with the wind, so the wind correction angle is zero and the inside wingtip points at the pylon.

- At 90 degrees of turn with a direct crosswind: Groundspeed and bank angle have decreased to intermediate values. The airplane is flying perpendicular to the wind, so the wind correction angle is greatest and the wingtip points behind the pylon.

- At 180 degrees of turn with a direct headwind: Groundspeed and bank angle have decreased to minimum values. The airplane is flying into the wind, so the wind correction angle is zero and the wingtip again points at the pylon.

- At 270 degrees of turn with a direct crosswind. Groundspeed and bank angle have increased back to intermediate values. The airplane is again flying perpendicular to the wind, so the wind correction angle is greatest but the wingtip now points ahead of the pylon.

The wing points at the pylon only twice during this maneuver.

Remember, the wingtip points at the pylon only twice during the maneuver—when the airplane is flying upwind and downwind. The maximum wind correction angle occurs only twice—when the airplane is flying crosswind. The steepest bank occurs at maximum groundspeed; the shallowest bank occurs at minimum groundspeed. The airplane's bank angle and wind correction angle are constantly changing.

Power
Slow cruise.

Trim
Trimmed for slow cruise flight at the entry. Since this maneuver is a prolonged turn, trim as required.

Vision
Use the wingtip and nose references to establish and maintain attitude. Pick out surface references that represent the desired ground track, and fly over these points. Continually cross-check airspeed, altitude, and airspace for traffic.

CFI Notes

Always plan ground reference training flights during periods when the wind is blowing. No training value exists when these maneuvers are performed in no-wind conditions.

When introducing turns around a point, pick a good grid reference with a well defined pylon. That makes it easy for the student to select the desired ground track. Students learn faster when they reference the desired ground track and the pylon as opposed to referencing only the pylon.

Do not advance students in their training until they fully understand and master this maneuver. Turns around a point are a prerequisite for the more advanced ground reference maneuvers. Planning a ground track and using bank angle to compensate for wind drift during turns are subtle concepts that must be mastered if a student expects to be proficient at flying a proper airport traffic pattern.

Lift is a powerful force, and large bank angle changes are seldom required when correcting for groundspeed variations.

When studying turns around a point, students are often misguided by the terms steepest and shallowest banks. These terms imply that the bank varies during the maneuver, not that it consists of steep and shallow bank attitudes. Lift is a powerful force, and large bank angle changes are seldom required when correcting for groundspeed variations.

EIGHTS AROUND PYLONS

At the time of publication, this excellent training maneuver was no longer required for private pilot certification, but that does not diminish its value or mean that it will not be reinstated in the future. My *Private Pilot Flight Training Syllabus* lists this maneuver as optional.

Purpose

To maintain a constant altitude and a specific ground track while flying a figure eight around two pylons. To recognize and compensate for wind drift. To plan and control circular and linear ground tracks while properly dividing your attention.

Execution

Clear the surrounding airspace for traffic. Pick out a road or other linear feature that is perpendicular to the wind. Find a well defined pylon that lies on the road and start a turn-around-a-point maneuver. When passing downwind of this pylon, parallel to the road, continue the turn for approximately

45 degrees and rollout to a wings-level attitude. When you cross the road, pick a second pylon on the opposite side of the airplane and perform another turn-around-a-point maneuver. Rollout when you are on a 45-degree intercept with the road and continue the maneuver.

Fly the maneuver at 800 to 1,000 feet AGL while using the principles for turns around a point and tracking a road—an imaginary line in this case. The turns around each pylon are in opposite directions and both radii must be equal.

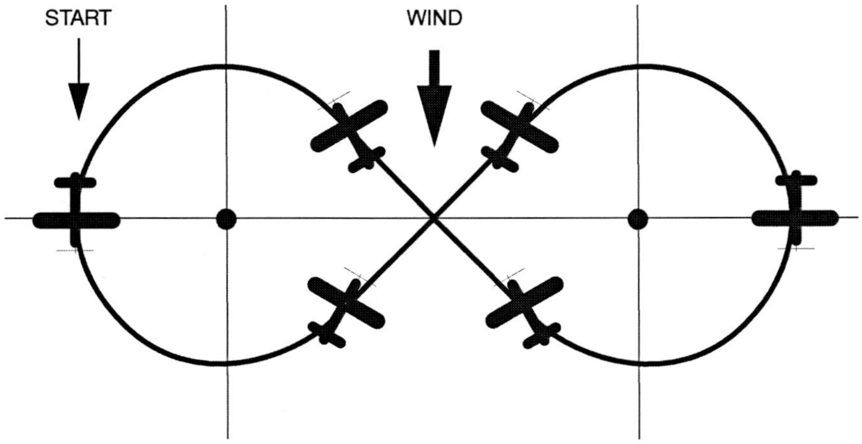

Eights Around Pylons

Attitudes
Same as for turns around a point.

Power
Slow cruise.

Trim
Trimmed for slow cruise flight at the entry.

Vision
Use the wingtip and nose references to establish and maintain attitude. Pick out surface references that represent the desired ground track, and fly over these points. Continually cross-check airspeed, altitude, and airspace.

ADVANCED TAKEOFFS AND LANDINGS

CROSSWIND TAKEOFF

Purpose
To compensate for a crosswind during takeoff.

Execution
Clear the surrounding airspace for traffic. As described by the Normal Takeoff and the Departure Profile, page 4 - 2, with the following adjustments:

Wind awareness You must make a point of continually monitoring *actual* wind conditions. This effort begins when you walk out to preflight the airplane, and it ends when you tie-down the airplane after your flight. Never assume that the actual wind will be equal to the wind that is reported by an air traffic controller, a unicom operator, ATIS, or an automatic weather reporting system—ASOS or AWOS.

You must look for actual wind indicators—wind socks, flags, blowing dust, or smoke.

You must look for actual wind indicators—wind socks, flags, blowing dust or snow, or smoke to include the smoke plumes that occur when a jet's tires contact the runway. Remember the saying, "To be forewarned is to be forearmed."

Elevator control For the crosswind takeoff, do not establish the T-pitch attitude as soon as the elevator becomes effective. Wait until elevator pressure is firm and control response is positive. This prevents a liftoff at minimum speed, and minimizes the chance of settling back onto the runway during gusty wind conditions.

Use rudder to keep the airplane parallel with the runway centerline until a positive climb occurs.

Rudder control Use it to keep the airplane parallel with the runway centerline until a positive climb occurs. You identify a positive climb by looking out the window at the pie-in-the-sky sight picture and evaluating airplane movement in relation to the ground.

The term *positive climb* does not mean *positive rate of climb.* Keep your eyes outside the cockpit and evaluate your climb performance visually. Instrument-rated pilots use the vertical speed indicator to confirm a positive rate of climb during instrument takeoffs in low visibility, instrument conditions.

Aileron control Use it to keep the airplane's weight on the upwind wheel during the takeoff roll and to maintain the liftoff bank attitude until a positive climb occurs.

To prepare for weight transfer in strong winds, you must apply full aileron displacement into the wind at the start of the takeoff roll. At some point during the acceleration, the downwind wing will start to rise. Smoothly decrease aileron input just enough to keep the downwind wing from rising appreciably. At liftoff, neutralize the ailerons, but maintain the slight bank angle that existed at liftoff.

When a positive climb occurs, simultaneously level the wings and apply enough right rudder to compensate for engine torque during the climb. The airplane will automatically yaw into the wind and establish a wind correction angle. Adjust that angle in order to track the runway's extended centerline. Use this procedure for all crosswind takeoffs.

The initial sideslip protects the airplane if it settles back onto the runway due to wind gusts.

The initial sideslip, required until a positive climb occurs, protects the airplane if it settles back onto the runway due to wind gusts. Touchdown in a sideslip is normal and easy to control; touchdown in a crab is extremely dangerous and can easily cause loss of control.

Never hold the airplane on the runway, accelerate to an excessively high speed, and then jerk the airplane into the air. This foolish action places too much weight on the airplane's nose wheel, a situation that will cause wheelbarrowing.

Control summary Aileron input, initially displaced into the wind, is reduced as the airplane accelerates in order to keep the downwind wing from rising appreciably. Rudder input is used to keep the airplane parallel with the runway centerline until a positive climb occurs. Elevator input, except for the delay to establish the T-pitch attitude, is identical with normal takeoff requirements.

The downwind wing rises slightly during the takeoff roll, the nose rises when the T-pitch attitude is selected, and the upwind main wheel is the last wheel to leave the ground. Terminate the sideslip when a positive climb occurs.

SHORT-FIELD TAKEOFF

During training, assume that a 50-foot obstacle exists.

Purpose
To takeoff and clear a 50-foot obstacle.

Execution
Clear the surrounding airspace for traffic. As described by the Normal Takeoff and the Departure Profile, page 4 - 2, with the following adjustments:

- Set flaps as recommended by the POH, taxi into position at the absolute end of the runway (not always possible during training), set brakes, set full power, release brakes, and begin the takeoff roll. Use the C- and T-pitch attitudes in the normal manner.

- When lift-off occurs, instead of maintaining the T-pitch attitude, smoothly establish the P-pitch attitude for best-angle climb, V_X. Stop increasing pitch if a stall warning or prestall buffet occur.

- When clear of the 50-foot obstacle, establish the T-pitch attitude, retract the flaps when the airplane starts to accelerate, and upon reaching V_Y, follow the departure profile.

SOFT-FIELD TAKEOFF

Purpose
To takeoff when departing from a soft-surface runway—standing water, wet or tall grass, snow, soft dirt, or mud.

Execution
Clear the surrounding airspace for traffic. As described by the Normal Takeoff and the Departure Profile, page 4 - 2, with the following adjustments:

Do not stop on the runway when making an actual soft-field takeoff.

- Set flaps as recommended by the POH, taxi onto the runway with the yoke held full aft, set full power, and begin the takeoff roll. Do not stop on the runway (not always possible during training). If you are performing an actual soft-field takeoff, higher than normal power will be required in order to taxi onto the runway, and if you stop, the airplane may become bogged down, unable to move forward.

- When pitch increases to the P-pitch attitude, *immediately* move the yoke forward and maintain that attitude until liftoff occurs. The stall warning will activate while you accelerate to the liftoff point. You must use the pie-in-the-sky sight picture during the entire soft-field takeoff procedure.

- At liftoff, decrease pitch attitude to avoid an excessive climb and stay close to the runway while the airplane accelerates. When you reach V_Y, establish the I-pitch attitude, retract the flaps, and follow the departure profile.

ABORTED TAKEOFF

Purpose
To terminate a takeoff.

Execution
You must maximize braking and avoid wheelbarrowing, particularly during a high-speed abort on a short runway. Here's the sequence: throttle closed, yoke full aft to maximize the weight on the main wheels for braking effectiveness, and brakes applied without skidding the tires. Do not release the yoke or report the abort to air traffic control until the airplane has decelerated to taxi speed.

ARRIVAL PROFILE

Purpose
To descend at 500 feet per minute from cruise altitude to traffic pattern altitude using the fastest forward speed. To minimize power and attitude changes. To optimize your ability to observe other traffic in the vicinity of the airport. To learn the importance of being a smooth, efficient pilot.

Execution
Clear the surrounding airspace for traffic. Start the descent at a point that will allow you to reach traffic pattern altitude four to five miles from the airport if terrain permits. Upon reaching pattern altitude, level off, set slow-cruise power, and let the airplane decelerate to slow-cruise airspeed as it approaches the downwind leg. When the airplane turns downwind it should be stabilized and trimmed for slow cruise. When ready to descend from pattern altitude, start CAAPVAA.

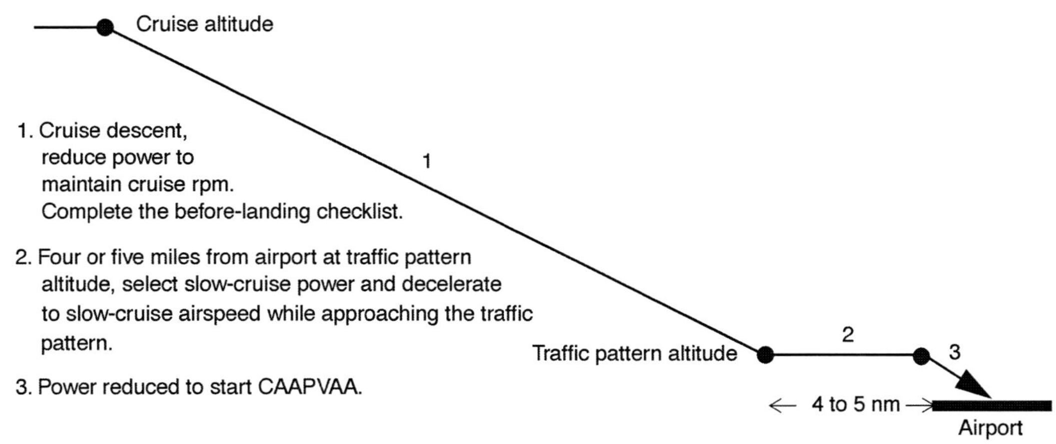

The Arrival Profile

A maximum of four power reductions are made between starting the descent and the landing.

Note that a maximum of four power reductions are made between starting the descent and landing: (1) A power reduction to maintain cruise rpm as you accelerate, to maintain airspeed limitations, or to reduce airspeed for turbulence. (2) A reduction to slow-cruise power when you level off at traffic pattern altitude. (3) A reduction to start CAAPVAA (4) A reduction to idle power during the landing flare. Flying in this manner will portray you as a smooth, efficient pilot.

This descent is normally flown at 500 feet per minute in unpressurized airplanes. If your passengers are accustomed to unpressurized flight, the descent rate can be increased.

Air traffic is easier to spot against a sky background rather than a surface background.

Plan the descent so that you avoid last-minute, high-speed, power-off approaches. Your goal is to end the cruise descent at pattern altitude, four to five miles from the airport, so that you can look for traffic against a sky background instead of the Earth's surface. Variations may be necessary due to local obstructions or minimum altitude requirements.

Here is the rule of thumb for computing the distance required to make a 500 foot per minute (fpm) descent:

DISTANCE TO DESCEND (in miles) AT 500 FEET PER MINUTE =

(ALTITUDE TO BE LOST) X (ESTIMATED GROUND SPEED) X (2)
in thousands of feet — in miles per minute

Example: You want to descend at 500 fpm from a cruise altitude of 8,500 to a pattern altitude of 1,000 feet with a groundspeed of approximately 120 knots or 2 miles per minute. The distance required for the descent will be 7.5 (altitude difference), times 2 (groundspeed), times 2, or 30 nautical miles. To reach pattern altitude five miles out, you start the descent 35 miles from the airport.

Here is the rule of thumb for computing the distance required to make a 1,000 fpm descent:

DISTANCE TO DESCEND (in miles) AT 1,000 FEET PER MINUTE =

(ALTITUDE TO BE LOST) X (ESTIMATED GROUND SPEED)
in thousands of feet in miles per minute

Example: You want to descend at 1,000 fpm from a cruise altitude of 8,500 to a pattern altitude of 1,000 feet with a groundspeed of approximately 180 knots or 3 miles per minute. The distance required for the descent will be 7.5 (altitude difference), times 3 (groundspeed), or approximately 22 nautical miles. To reach pattern altitude five miles out, you start the descent approximately 27 miles from the airport.

CROSSWIND APPROACH AND LANDING

Purpose

To compensate for a crosswind during approach, landing, and roll-out.

Execution

Clear the surrounding airspace for traffic. When you turn onto final approach, always rollout on and parallel to the extended runway centerline so that you can evaluate wind drift. If drift occurs, stay on the centerline by crabbing into the wind until it is necessary establish a sideslip for landing. That transition should occur no later than 100 to 200 feet above the runway.

Never touchdown in a crab, an action that can damage the landing gear and cause loss of control. Large airplanes can transition from the crab to the sideslip as they flare, because their large mass (inertia) momentarily nullifies some of the wind effect. Light airplanes do not have this advantage.

Remain parallel with the centerline with rudder; stay over the centerline with ailerons.

The sideslip is easily established and maintained by using a simple relationship: Remain parallel to the centerline with rudder; stay over the centerline with ailerons. This technique automatically crosses the controls and establishes the slip.

If the winds are gusty, increase your normal approach speed for controllability. For light airplanes, add one-half the gust factor up to a maximum correction of 10 knots. The higher airspeed gives you more control authority.

At touchdown, increase aileron input and glue the upwind wheel onto the runway.

When the airplane touches down on the upwind main wheel, immediately increase aileron and elevator displacement in order to maintain the existing pitch and bank attitudes for as long as possible. This procedure keeps the nose wheel and downwind main wheel from touching down prematurely, and it transfers the most weight to the upwind main wheel, a critical requirement for eliminating drift. Think of this as gluing the upwind wheel onto the runway. The downwind main wheel and the nose wheel will touch down automatically as airspeed decreases. Here are the control actions that occur during a well executed crosswind landing from a sideslip:

Rudder control Use it to maintain heading and keep the airplane parallel with the runway's centerline. When the nose wheel touches down, momentarily center the rudder pedals to prevent swerving as the nose strut compresses, and then re-apply rudder for directional control.

In most airplanes with nose wheel steering, strut compression engages a cam that aligns the nose wheel with the rudder pedal position. When the nose strut is extended, the nose wheel is mechanically aligned with the airplane's longitudinal axis in order to minimize drag.

Aileron control Use it to control bank and keep the airplane over the runway centerline. At touchdown quickly increase aileron displacement in order to maintain the touchdown bank attitude and place the maximum amount of weight on the upwind main wheel. Apply full aileron displacement as rapidly as possible. In strong winds, the aileron position at the end of a crosswind landing will be the same as it was when you started the crosswind takeoff.

Elevator control Use it to establish the landing pitch attitude. At touchdown, increase elevator pressure somewhat in order to counteract the pitch down induced by the upwind main wheel contacting the runway, and maintain that attitude until the pitch attitude automatically decreases.

Control summary Just after touchdown on the upwind main wheel with the yoke well aft, you displace the ailerons further into the wind and use the rudder for directional control. The downwind main wheel will touch down when the ailerons lose their aerodynamic effectiveness; the nose wheel will touch down when the elevator loses its aerodynamic effectiveness. Crosswind landings are a one, two, three touchdown sequence—upwind main wheel, downwind main wheel, and finally the nose wheel.

The absolute bank limit for crosswind landings is the amount of bank that you can establish while keeping the airplane parallel to the runway centerline with rudder. You should always observe the maximum allowable crosswind component if one is published in the POH.

If a safe crosswind landing is questionable, divert to an alternate airport.

Pilots of low-wing airplanes are usually concerned that a wingtip will strike the ground during strong-crosswind landings, but the airplane will normally run out of rudder before that occurs. If you attempt a crosswind landing with a bank angle that requires maximum rudder deflection, a safe landing, particularly in gusty wind conditions, is very doubtful and very dangerous. Be safe, divert to an alternate airport.

The crosswind component can be reduced somewhat by landing diagonally across the centerline of a wide runway. Plan to touch down near the runway's downwind edge, and complete the rollout along the runway's upwind edge.

In extremely strong winds, call for a wing-walker before taxiing.

Use extreme caution when landing in extremely strong winds. Many pilots have landed successfully in these conditions only to lose control of the airplane while taxiing to the ramp. If those conditions exist, turn into the wind after landing, taxi clear of the runway, and call for a *wing walker.* That person should hold onto the upwind wingtip while you slowly taxi to the tie-down spot.

SPECIALIZED APPROACHES AND LANDINGS

Think of short-field, soft-field, and no-flap approaches and landings as specialized procedures.

Short-field, soft-field, and no-flap approaches and landings—explained next—are specialized rites of passage for obtaining a pilot's certificate. I say specialized, because to think of these seldom used, advanced procedures as just a variation of the CAAPVAA and normal landing is a mistake.

Here's the trick that makes both procedures easy to execute whenever one is required: Make both approaches identical and use power for glidepath control. Plan your traffic pattern so that you will be above the proper glidepath when you turn onto final approach. At that point, immediately select idle power and the C-pitch attitude and extend full flaps.

The limited wing flaps of most modern-day, light airplanes cannot be extended far enough (at least 40 degrees) to create the drag that is required for a steep, short-field approach, and the amount of power that you must use is a compromise. Idle power would result in the steepest glidepath, but a slight amount of power is necessary so that you can compensate for updrafts during the approach.

Continue the power-off descent until it becomes obvious that power must be added in order to reach the runway. Add power to capture that glidepath, but do not look at the tachometer for that adjustment. Just make an audible power increase and adjust pitch attitude slightly to establish the correct airspeed. Now use the visual runway references to evaluate the need for power, and maintain your pitch attitude until you start the landing flare. Change power as necessary in order to stay on the glidepath.

Thrust is a powerful force. In benign conditions small power changes and a constant pitch attitude will compensate for minor glidepath deviations without causing significant airspeed changes. However, if conditions are gusty or up-and-down drafts (risers and sinkers) occur, large power changes will be required, and you must then change pitch attitude in order to maintain airspeed. Throttle movement will be almost continuous in extreme cases with moderate to severe turbulence.

For training, there is only one difference between short- and soft-field landings: Close the throttle when you start the short-field landing flare; close it after the soft-field landing touch-down.

For pilot certification, soft-field landings are not accuracy landings, but in actual practice they are. Assume that you are approaching a waterlogged grass strip, but on final approach you see an area that looks relatively dry. That is definitely where you want to land, and that decision mandates a soft-field accuracy landing.

After an actual soft-field landing, do not close the throttle until after you assess your ability to taxi. A power increase is usually required in order to keep moving. If you close the throttle and let the airplane stop rolling, it may settle into the soft ground. Now you are stuck, and you may be unable to taxi.

Short-field landings have special considerations as well, but these occur before you start the approach. Whenever you fly to a short, unfamiliar field, you should conduct both a high and low reconnaissance of the landing site.

First, you should circle the field at 800 to 1,000 feet agl in order to look for obstructions such as wires and to evaluate the surrounding area for your subsequent departure. If possible, your departure direction should give you the best forced landing option.

Next, descend to 100 to 300 feet above the highest obstacle, and fly parallel to and close enough to the landing site so that you can evaluate surface conditions and slope. Pick out a touchdown point and a go-around point using readily identifiable ground references. Climb back to pattern altitude and fly your approach. If you're not on the ground when you reach the go-around point, go-around and readjust your approach.

When a large obstacle is involved, particularly a tree line or other linear obstruction, never close the throttle and decrease pitch attitude (dump the nose) when you clear the obstacle. Your short-field glidepath should be a straight-line descent all the way down final approach to the touchdown point, and that line should just clear the obstacle.

The wind gradient will usually change when you descend below the obstacle, particularly if it is a linear obstruction. If you set idle power, dump the nose, and induce a high sink rate just as you clear the obstacle, you may not have enough elevator authority for the landing flare. That situation has caused many short-field landing accidents. In many cases, you must add power when you descend below the obstacle.

For accuracy landings, you must fly all approaches at the correct airspeed. The manufacturer's published airspeed is for an airplane at maximum gross weight. For every two percent reduction in gross weight, you can reduce the approach speed by one percent. This is an important rule of thumb to follow when an accuracy landing is required. If gusty winds exist, you should increase the corrected airspeed by one-half the gust factor, not to exceed a 10 knot increase.

SHORT-FIELD APPROACH AND LANDING

Purpose
To land and stop the airplane in the shortest possible distance.

Execution
Clear the surrounding airspace for traffic. As described in the previous section, Specialized Approaches and Landings, and the sight pictures on pages 4 - 15, 16, and 17.

At touchdown apply light braking and move the yoke aft as rapidly as practical in order to maximize the weight on the main wheels. Apply heavier braking when your speed decreases significantly.

Maximum wheel weight, required for maximum stopping effort, can be obtained by retracting the flaps at touchdown. This action, however, violates the SOP that prohibits moving anything but the flight controls while on the runway.

A maximum effort stop is not a requirement, because takeoff distance is always greater than landing distance—you would never land at a site that you could not depart from. When performing short-field landings, always consider practicality, safety, and airplane wear and tear. Use common sense.

SOFT-FIELD APPROACH AND LANDING

Purpose
To land on soft ground at the slowest possible speed. To keep the weight off the nose wheel as long as possible.

Execution
Clear the surrounding airspace for traffic. As described for short-field landings except that a slight amount of power is maintained until touchdown occurs.

NO-FLAP APPROACH AND LANDING

Purpose
To land an airplane without the use of flaps.

Execution
Clear the surrounding airspace for traffic. As described in the previous section, Specialized Approaches and Landings, and the sight pictures on pages 4 - 15, 16, and 17 with the following adjustments.

- Without flap drag, the descent angle decreases and you must fly a larger traffic pattern. Fly a wide downwind leg or fly a normal downwind leg but make the base leg turn well past the normal, 45-degree reference point.

- When ready to descend from traffic pattern altitude, set idle power and establish the T-pitch attitude. This slightly nose-high attitude should generate the proper approach speed. Add power when a shallower-than-normal glidepath is intercepted, and use power to maintain that glidepath.

- No-flap approaches obviously have much less drag than flaps-extended approaches. This changes the effect of power corrections. A slight power increase will be very effective; a slight power decrease will not. Therefore, you must anticipate the need for a power reduction and apply it much earlier than you would with flaps extended.

A landing flare is seldom required for a properly flown no-flap approach.

- Set idle power just prior to touchdown, and keep the yoke aft after touchdown in order to avoid wheelbarrowing. A landing flare is seldom required for a properly flown no-flap approach.

EMERGENCY PROCEDURES

BEST-GLIDE DESCENT

Purpose
To maximize the distance traveled when gliding without power in calm wind.

Attitude
Establish and maintain the C-pitch and B-bank attitudes.

Power
None or idle, propeller windmilling.

Trim
When airspeed stabilizes, fine-trim the elevator.

Vision
Clear the surrounding airspace for traffic. Use the wing and nose positions to set and maintain attitude. Use the Earth's surface and the horizon, framed by the nose cowl and the windshield's framework, to evaluate roll, pitch, and yaw.

Glide distance increases with a tailwind and decreases with a headwind. The manufacturer's best glide speed is for an airplane at maximum gross weight flying in calm wind. Best glide speed decreases as weight decreases and when flying with a tailwind. It increases when flying into a headwind.

MINIMUM-SINK DESCENT

Purpose
To minimize the descent rate and maximize flight time when gliding over or near a suitable emergency landing site.

Attitude
Establish and maintain the T-pitch and the B-bank attitude, or if you must circle the landing site, use the-A bank attitude.

Power
None or idle, propeller windmilling.

Trim
When airspeed stabilizes, fine-trim the elevator.

Vision

Clear the surrounding airspace for traffic. Use the wing and nose positions to set and maintain attitude. Use the Earth's surface and the horizon, framed by the nose cowl and the windshield's framework, to evaluate roll, pitch, and yaw.

EMERGENCY DESCENT

Purpose

To minimize flight time when you descend due to an engine or uncontrollable cabin fire. This is a dire emergency.

Attitude

Establish the B-, A-, N-, or K-bank attitude in order to reach a suitable landing site. As bank increases, vertical speed increases—the K-bank attitude produces the fastest descent rate. Adjust pitch attitude to achieve the highest possible airspeed: V_{NE} in smooth air, V_{NO} in light to moderate turbulence, and V_A in severe turbulence.

Power

None or idle, propeller windmilling.

Trim

When airspeed stabilizes, fine-trim the elevator.

Vision

Clear the surrounding airspace for traffic. Use the wing and nose positions to set and maintain attitude. Use the Earth's surface and the horizon, framed by the nose cowl and the windshield's framework, to evaluate roll, pitch, and yaw.

EMERGENCY APPROACH AND LANDING

Purpose

To properly manage the airplane after an engine failure. To make a safe, power-off approach and landing.

Execution

Clear the surrounding airspace for traffic. Follow the ABC checklist: **A**irplane, **B**est landing site, and **C**ockpit. The philosophy of this checklist is survival—keep the airplane flying, select a suitable landing site, and attempt an engine restart.

If the engine quits, establish the C-pitch attitude and pick a landing site before you look inside the cockpit.

Airplane Look outside the cockpit and establish the C-pitch, best-glide attitude. *Do not look inside the cockpit!* It takes time for the airplane to decelerate to the best glide speed, and this time should be used for the next step—selecting a suitable landing site.

Best landing site Do this while the airplane decelerates and you trim for the C-pitch attitude. Otherwise, you will waste precious time.

If altitude is extremely low, your only option is to pick an area of least resistance and land straight ahead. No altitude is available for maneuvering.

The area of least resistance means a landing site that will result in minimum fuselage damage—a site that will damage the wings but not the fuselage. Two trees separated by at least the width of the fuselage or a football goal post are examples. Many pilots have walked away from impossible situations because they protected the fuselage and let the airplane's wings take the punishment.

If somewhat higher, you may be able to turn up to 90 degrees. Your highest priority is an upwind landing with minimum groundspeed; your lowest priority is a downwind landing with high groundspeed. If you hit an obstacle on landing, the amount of damage is directly related to groundspeed.

If at or somewhat above traffic pattern altitude—about 1,000 feet agl—with no visible options, turn downwind and look for an area of least resistance on either side of the airplane. After turning downwind, you should have enough altitude to complete two 90-degree turns and land into the wind.

At higher altitudes, you should maintain the best-glide attitude and roll the airplane into brief left and right steep bank turns. While banked, start looking directly beneath you for a suitable landing site, and move your field of vision outward from that point if need be. Hopefully, you will find an area that is close to your position so that you have more time to plan your approach and attempt the engine restart. As previously mentioned, if a good area is not evident, turn downwind in order to maximize groundspeed, glide distance, and landing options.

Once you have chosen a landing site, keep the touchdown spot in sight, and do not change your mind.

When you find a suitable site, turn toward it, and attempt the engine restart procedure. If the engine will not start, continue toward the landing site and, if altitude permits, spiral down over it so that you never lose sight of your planned touchdown point. Roll out on the downwind, base, or final leg to complete the landing. Never change your landing site decision unless a far better solution becomes readily apparent.

Turn the fuel selector and master switch off, and open the cabin door to avoid a jammed door in case of structural failure during the landing. Expect a fixed-gear airplane to flip over if the landing is made on extremely soft ground or water.

Cockpit Use the cockpit flow pattern to attempt the engine restart. Engine failure is the real reason for teaching flow patterns. If you are at low altitude, there is no time to consult a written checklist, and your performance must be flawless.

Do the following if applicable: electric fuel pump on, switch fuel tanks, mixture set for altitude, carburetor heat on, and primer locked.

If the engine starts running for no apparent reason, use caution. It may quit again just after you abandon your approach.

If the engine starts running again for no apparent reason, use caution. It may quit again just after you abandon your approach.

Sudden engine stoppage is normally caused by fuel starvation—a failed engine fuel pump, an empty fuel tank, an excessively leaned mixture, or a block fuel tank vent. Gradual power failure is normally caused by carburetor ice, a blocked air filter, or a leaned mixture that was not reset during a descent.

Engine roughness is normally caused by bad magneto timing, a rich mixture that occurs when you leave the carburetor heat on in order to prevent carburetor ice, or engine structural failure. If magneto timing is the problem, isolate each magneto to determine the problem. If the roughness disappears, leave the magneto switch in that position and land as soon as practical.

ATC Radio Communications If you are using ATC's radar advisory service when engine failure occurs, consider these facts:

- You must not have any interruptions until you have completed the ABC mental checklist. Never call ATC until this objective has been completed.

- When you do report the engine failure and squawk the emergency transponder code, most controllers will start asking you several questions.

- If time is short, you *must* ignore those questions and concentrate on making a safe landing so that you and your passengers survive uninjured.

- If time is plentiful, you can talk, but keep it short.

- ATC communications are your last priority.

CFI Notes

Students frequently ask if stopping propeller rotation will improve glide performance. No, because airspeed must be reduced close to the stall speed and then increased back to best glide speed, an action that grossly degrades glide performance and will seldom stop the propeller.

If propeller rotation was stopped due to catastrophic engine failure, glide distance would increase. I would attempt to stop propeller rotation only if altitude was sufficient and the failed but rotating engine was causing severe airframe vibrations.

225-DEGREE TURN AFTER ENGINE FAILURE

Purpose

To simulate a power failure and an immediate turn back to the departure runway, a turn that could be well over 180 degrees. To perform a 225-degree turn with minimum loss of altitude after engine failure during a best-rate climb. To determine the altitude loss for this maneuver.

Execution

Clear the surrounding airspace for traffic. Simulate engine failure during a best-rate climb and note the altitude. Simultaneously establish the best-glide pitch attitude and roll the airplane into a steep-bank turn. Roll out after 225 degrees of turn and note the altitude loss.

The actual no-wind turn for runway alignment could require a 225-degree turn in one direction followed by a 45-degree turn in the opposite direction. If a crosswind exists, you should turn into the wind in order to reduce the amount of turn and minimize the distance to the runway. A downwind turn would increase that distance. The altitude loss that you determine could almost double due to your reaction time and skill, aircraft weight, and additional heading changes that may be required for runway alignment during crosswind conditions.

Attitude
To start the maneuver, establish the I-pitch and the B-bank attitudes. When engine failure occurs, simultaneously establish the K-bank and the H-pitch attitudes. During the rollout from the 225-degree turn, establish the C-pitch attitude if you need best-glide performance prior to flap extension.

Power
Idle power to simulate engine failure.

Trim
Have the airplane trimmed for the climb. Retrim for the glide after the 225-degree turn has been completed.

Vision
Clear the surrounding airspace for traffic. Use the wing and nose positions to set and maintain attitude. Use the Earth's surface and the horizon, framed by the nose cowl and the windshield's framework, to evaluate roll, pitch, and yaw. Note the altitude that was lost from the time of engine failure until the completion of the 225-degree turn.

180-DEGREE, POWER-OFF APPROACH AND LANDING

Purpose
To perform a 180-degree, power-off approach and landing when located on the traffic pattern's downwind leg and abeam the planned touchdown point.

Execution
Clear the surrounding airspace for traffic. Use CAAPVAA procedures (no power is constant power) and the three stages of a good landing with the following adjustment.

Think of fishing when you perform this maneuver. A big fish takes your bait, you set the hook, and you then play the fish in and out until it tires to the point where you can reel it in.

The runway is the fish, and you set the hook by turning toward the runway (or forced landing site) when engine failure is simulated. Starting the turn, you immediately realize that you are too high, so you turn away from the runway and extend the first increment of flaps.

After a few moments you should again turn toward the runway. When you sense that you are too high, turn away from the runway. Continue to do this until you turn onto final approach. These are small turns that keep you just on the edge of that "I'm too high" sensation—you are playing the fish in and out.

This is no time for smooth flying. Be aggressive with the airplane. Don't dawdle. You do not want the fish to get away—you get too high—nor do you want it to break your line—you get too low. On base leg extend the second increment of flaps.

On final approach delay full flap extension for a moment, even though you may seem high. You have just turned into a headwind, which will decrease your groundspeed and increase your descent angle. Evaluate the wind effect and then make your decision on full flap extension. By staying just on the edge of the "I'm too high" sensation, you still have one tool left that will ensure success—the forward slip.

PRECAUTIONARY LANDINGS

Purpose

To intentionally land an airplane at a suitable, non-airport site. To realize that a planned, power-on landing is far better than continuing a flight into a hazardous situation or encountering a high-risk, power-off emergency landing. A hazardous situation implies getting lost, low fuel, deteriorating weather, engine problems, or nightfall.

Execution

Clear the surrounding airspace for traffic. Pick a suitable landing area such as a road or a hard-surfaced field. Try to locate a site where shelter, a telephone, or transportation might be nearby. Land as close as possible to that site, and keep its location oriented in relation to the landing area. Everything will look different when you start walking.

Perform a high and low reconnaissance of the landing site as described for the short-field approach and landing. Plan to make a short-field approach and landing.

Prior to landing, use the emergency frequency, 121.5 MHz, and transmit your situation in the blind. Low altitude radio facility charts depict Air Route Traffic Control Center frequencies that are used by pilots on instrument flight plans. If you have these charts in your possession, radio contact with overflying aircraft is possible.

INSIGHTS (One of my articles from the AOPA's *Flight Training* magazine)

Rock And Roll

Dancing with an airplane - May 2004 issue

Late last year, my airport experienced strong, gusty crosswind conditions that generated occasionally severe, low-level turbulence. I told my stage-check students that they could reschedule, but both wanted to give it a try. That's the spirit! Never pass up an opportunity to gain experience.

I told them to stay in the traffic pattern in order to experience airplane-style rock and roll. You will benefit from what they learned.

You cannot just sit in the pilot's seat, wiggle the yoke, and hope that everything will work out. You must treat the airplane just like mother nature is treating it—if the air is smooth, you can be smooth; if the air if rough, you must be rough.

Rough in this case means that you must be very aggressive with the flight controls in order to keep the airplane's attitude exactly where you want it. During the landing approach, airspeed will quickly increase or decrease, and you cannot correct it with pitch due to your low altitude. You must be aggressive with the throttle and go after airspeed with power.

Do not make assumptions as to what will happen next. Just be ready for the extreme variations that will occur. Len Morgan, a retired Braniff Airlines captain, once wrote, "Flying is just like playing cards. You always play with the same deck, but you never get the same hand twice." When it comes to flying airplanes, nothing says it better than that.

If you were flying with me in these conditions in an airplane that had dual control sticks and I told you hold your stick and follow my actions, you would feel the stick continually move in a small square pattern that I call the *feel box*. I'm constantly assessing attitude response to control inputs, and when needed I make momentary, aggressive pressure inputs to maintain attitude. When significant airspeed variations occur, I make momentary, aggressive power changes.

Two-step corrections are mandatory. When attitude or airspeed start to deviate, apply a correction that you know is excessive; when that correction starts to work, immediately remove half of it.

A slow, gentle landing flare from 30 to 40 feet agl doesn't cut it. Mother nature will eat your lunch. You must force the airplane down so that it's close to the runway surface, chop the power, and forcibly work the airplane into the landing attitude prior to touchdown. If you do not shift your eyes to the left or right of the nose cowl during the landing flare (the right side is used when slipping into a right crosswind), you will very likely experience what surfers call a "wipe out." You compromised your safety because you could not see the drift and sink rate variations that mandate immediate corrections, and worst of all, you probably landed flat or nosewheel first (extremely dangerous) instead of landing with the proper nose-high attitude so that the nose wheel is well clear of the runway.

For a crosswind landing, do not fall into the trap of mechanical thinking—over analyzing what you are doing. Yes, there are many reasons that explain why airplanes react the way they do, and that's important knowledge for ground school. To nullify those factors when landing, use rudder to keep the airplane's longitudinal axis parallel with the runway centerline; use aileron (bank angle) to keep the airplane over the runway centerline. That's a sideslip, but who cares. Just fly the airplane and put it where it belongs.

You must increase airspeed in order to maintain flight control effectiveness. For light airplanes, add at least half the gust factor to your normal final approach speed. You certainly don't want to be any slower then that, and there's nothing wrong with being a little faster if it gives you more control effectiveness.

Once you obtain the wind from the tower, do not ask for subsequent wind checks. Wind is too dynamic for that information to be helpful. What you sense is what you have, and it will change every second. Wind sensors can't keep up in this situation.

And always remember this: If you are not using the rudder aggressively, you are not dancing with the airplane. Don't be a wheel-wiggler and a pussyfooter. Use the rudder.

REVIEW TEST

1. What is the most important cockpit instrument?
2. What are the two principle magnetic compass errors?
3. What does best angle climb (Vx) accomplish?
4. What three mechanical corrections must you apply during steep turns?
5. How do you correct minor altitude variations during a coordinated steep turn?
6. Can you control an airplane if it is kept in a stalled condition with full aft elevator?
7. Can you stall an airplane at any airspeed?
8. What could happen if you inadvertently slip or skid while turning onto final approach?
9. What is the basic spin recovery technique?
10. Can you spin an airplane if it is not stalled?
11. Where do the steepest and shallowest banks occur during turns around a point?
12. Where does the maximum crab angle occur during turns around a point?
13. Why should you always start ground reference maneuvers while flying downwind?
14. Should you retrim the elevator during turns around a point?
15. When taxiing out for takeoff, what must you constantly be aware of?
16. How does the control application for crosswind takeoffs differ from normal takeoffs?
17. How does the control application for crosswind landings differ from normal landings?
18. If you experience difficulty during a crosswind landing, what is the most important thing to avoid?
19. What does a positive climb mean?
20. Summarize the short-field takeoff pitch-attitude selections.
21. Summarize the soft-field takeoff pitch-attitude selections.
22. What is the purpose of the arrival profile?
23. If your descent ground speed will be two miles-per-minute and you are 7,500 feet above traffic pattern altitude, how far from the airport would you start a 500 foot-per-minute descent?
24. How do you use the ailerons during a crosswind landing?
25. How do you use the rudder during a crosswind landing?
26. Describe the final-approach glidepath for a short-field landing.
27. What is the difference between best-glide and minimum-sink descents?
28. What is the mental checklist for engine failure?
29. What must you know before you would attempt a 180-degree turn back to the airport after engine failure?
30. What option do you have when your safety becomes compromised due to low fuel or degrading weather conditions?

CHAPTER 6

BASIC CROSS-COUNTRY NAVIGATION AND FLIGHT PROCEDURES

BASIC NAVIGATION PROCEDURES

Primary Considerations

Aviation safety consists of three elements—attitude, proficiency, and knowledge.

"Aviation safety is really made up of only three elements: attitude, proficiency, and knowledge"—R. C. Morton, FAA Safety Program Manger, retired. How true! One aspect of attitude is pessimism, a trait that successful pilots display whenever they prepare for and conduct a flight.

These pilots always plan for a worst-case scenario; they always ensure that an *out* exists—a alternate course of action that will guarantee their safety if something goes wrong. This is the ultimate aviation insurance policy, and the price is right. It's free!

Basic navigation training consists of navigation charts and pilotage, land-fall, and dead-reckoning navigation—your *out* when radio or satellite navigation is unavailable due to failure, maintenance, altitude, or location. You would be grossly negligent if you did not master and maintain these skills.

Ground Training

You can accomplish this essential training on your own by using commercial books or video programs or by attending a formal ground school. In either case, your flight instructor will spend considerable time with you to ensure that you know how to read navigation charts, plot courses, complete a navigation log, obtain and evaluate weather reports, utilize in-flight resources, determine aircraft performance, complete the aircraft's weight and balance computations, and file a flight plan with air traffic control (ATC).

Learning to fly is not like driver's education.

Please remember that to obtain a pilot certificate, your flight instructor must sign an FAA form and attest to your competency. No flight instructor will jeopardize his or her standing with the FAA and sign that form until you prove that you have mastered every requirement. This is not like drivers education. You are training to be the pilot in command of an airplane. That position carries tremendous responsibility, and it requires extensive ground and flight training.

PILOTAGE NAVIGATION

You follow well defined features that are depicted on navigation charts—highways, railroad tracks, and rivers are examples. Pilotage is not efficient because it seldom provides a direct route between two points, but it is very reliable. In desert or mountainous terrain where good surface references are difficult to find, pilotage may be the only way to navigate with confidence.

LANDFALL NAVIGATION

Pilotage and landfall navigation can eliminate confusion.

You fly an offset course to a prominent geographical reference that will then lead you to your objective—an airport or an en route reference point. This eliminates the direction-of-turn dilemma that can easily occur if you attempt to fly direct to your objective.

Consider a flight from Kansas City to Saint Louis. An easterly course could be flown direct to Saint Louis which is located on the Mississippi River, but if the you reached that impossible-to-miss river and did not see Saint Louis, an dilemma exists. Do you turn right or left?

To use landfall navigation, you fly a course that will definitely place you north or south of Saint Louis. Upon reaching the river, you will then know which way to turn, and finding Saint Louis is guaranteed.

DEAD-RECKONING NAVIGATION

You use plotted courses and computed headings, speeds, and times. (The word *dead* is derived from *deduced*.) This is the most efficient form of non-radio navigation because it permits a direct flight between any two points.

Route Selection

To fly with outdated charts is just asking for trouble.

You must clearly depict your route of flight on your navigation chart. (A chart is a map that you draw courses on.) Use a soft-lead pencil so that you can make a readily identifiable line that can be erased in order to reuse the chart. Terminal and sectional charts are normally published every six months; to fly with outdated charts is just asking for trouble.

Your initial objective for route selection is a direct flight from your departure airport to your destination airport. Obviously, that is not always possible due to terrain, airspace restrictions, weather, the lack of ground reference points, forced landing considerations, or your personal desires. Each leg on your route should begin and end at a reference point that is easily recognized from the air.

Start and end your route at the center of the airport symbols on your navigation chart.

Your route should start at the center of the departure airport symbol on the navigation chart and end at the center of the destination airport symbol. Never try to second guess what ATC will do to you when operating at controlled airports. As a pilot, you must always expect the unexpected; it happens on a regular basis.

After you plot the route on your chart, look for and circle easily identifiable reference points that lie on or adjacent to that route. One such reference every ten miles or so is sufficient, but use care when you make your selections. Many manmade references that look excellent on the chart are hard to see from the air, while many natural references that seem insignificant on the chart are easy to see.

The higher you fly, the harder it is to find some reference points. It becomes much easier if you keep your altitude within 3,000 feet of the Earth's surface.

Time is the real secret for finding the reference points that you circled on your chart.

Time is the real secret for finding reference points, but these times are not written on your navigation log. They are rough estimates based on your approximate groundspeed and estimated distance using the techniques that are explained in the Inflight Procedures section of this chapter.

After your reference points have been circled, you can, if desired, draw tic marks at ten-mile intervals along your entire route of flight. En route orientation is the reason you circle reference points and draw the ten-mile tic marks.

The objective of dead-reckoning navigation is not to fly from one reference point to the next. The object is to stay on your course line and use the reference points to identify your position and correct for wind drift. This is the secret for extremely accurate dead-reckoning navigation.

Wind is a variable that you must constantly deal with, because it changes as altitude, terrain, and atmospheric conditions change. The forecast wind that you base your flight on is a weather forecaster's estimate. Learn to be wind-wise, and fly headings that keep you directly over your course line. Staying close to the course line is not good enough.

Finally, draw a course arrow next to each leg for your direction of flight, and list the *magnetic* course and the leg distance as shown in the following illustration.

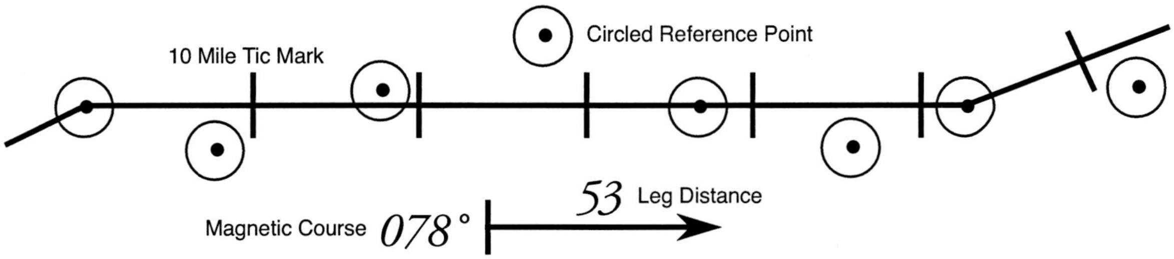

Navigation Leg and Course Arrow

The Navigation Log

Next, you complete a navigation log. Most commercially published logs have columns labeled checkpoints, true course, wind correction angle, true heading, variation, magnetic heading, deviation, compass heading, estimated and actual en route and arrival times, and fuel required.

A column for magnetic course rarely appears on navigation logs.

It is important to note that a column for magnetic course rarely appears on these logs, yet you must know it in order to determine the proper cruise altitude. That's why you compute the magnetic course and list it on the course arrow.

You list only the end points of each leg on your navigation log; consequently, those reference points now become checkpoints. When time permits during the flight, you calculate your groundspeed for that leg, update the estimated arrival time at subsequent checkpoints, and confirm your fuel consumption. There are two exceptions to this rule:

- On the first leg, starting after the point where you expect to be at cruise altitude (top of climb), you want two reference points listed on the navigation log that are at least 20 miles apart so that you can make a groundspeed check as soon as possible in order to confirm the forecast winds aloft. If your first leg is not extremely long, the second ground-speed reference point may be the end point for that leg.

You must make a groundspeed check early in your flight in order to validate the winds aloft and your fuel reserves.

This groundspeed check is critical, because if it's slower than planned, you must recalculate both your estimated time of arrival at your destination and your fuel reserves. It may be necessary to make an unplanned, en route fuel stop.

- On the last leg, you want one reference point listed on the navigation log that is at or near the point where you plan to start your descent (top of descent).

The previously mentioned navigation terms are listed next. You must have a full understanding of the following terms in order to complete the navigation log:

- True course—Your course based on true north. You determine this by using a navigation plotter to measure the angle between your plotted course and true north.

• Wind correction angle or drift angle—You determine this angle using a mechanical or electronic (recommended) flight computer. Whenever you need to change the word *course* to *heading*, you use the wind correction rule: Add the wind correction angle if you have a right crosswind; subtract it if you have a left crosswind.

• True heading—Your heading relative to true north. It is your true course ± the wind correction angle

Magnetic variation is depicted on navigation charts.

• Magnetic variation—The angle between the magnetic and true north poles. It is depicted on your navigation chart, as easterly or westerly variation. Whenever you need to change the word *true* to *magnetic*, you use the magnetic variation rule: "East is least (subtract), west is best (add)."

• Magnetic heading—Your heading relative to magnetic north. It is your true heading ± magnetic variation.

Compass deviation is listed on the airplane's magnetic compass housing.

• Compass deviation—The compass errors that exist due to magnetic fields within the airplane. It is listed in the airplane on a placard that is usually attached to the magnetic compass housing. Whenever you need to change the word *magnetic* to *compass,* you use the compass deviation card.

• Compass heading—The magnetic heading corrected for compass errors as listed on the compass deviation card.

You must have a full understanding of the following terms in order to determine magnetic course for the course arrow:

• True course—Your course based on true north. You determine this by using a navigation plotter to measure the angle between your plotted course and true north.

• Magnetic variation—The angle between the magnetic and true north poles. It is depicted on your navigation chart, as easterly or westerly variation. Whenever you need to change the word *true* to *magnetic*, you use the variation rule: East is least (subtract), west is best (add).

• Magnetic course—Your course based on magnetic north. It is your true course ± magnetic variation.

The following table summarizes the previous navigation terms and their relationships.

NAVIGATION LOG PREPARATION	COURSE ARROW PREPARATION
True Course	True Course
± Wind Correction Angle =	± Magnetic Variation =
True Heading	Magnetic Course *
± Magnetic Variation =	** If you fly more than 3,000 feet above ground level, you must fly at odd thousands plus 500' if your magnetic course is from 000° to 179° (e.g., 3,500, 5,500), and fly at even thousands plus 500' if your magnetic course is from 180° to 359° (e.g., 4,500, 6,500).*
Magnetic Heading	
± Compass Deviation =	
Compass Heading	

AVIATION WEATHER

You must understand basic meteorology and be able to interpret official aviation weather information.

Meteorology is a science, and while you are not expected to be a meteorologist, you must understand basic meteorology and be able to interpret official aviation weather information. Two government publications are required for this task: The current versions of Advisory Circular 00-6, *Aviation Weather*; and Advisory Circular 00-45, *Aviation Weather Services*.

If you attempt to study these manuals when you start flight training, you will probably become bored and lose interest. So, save them for later. You are not required to master this subject until you are in the last stage of flight training.

To make meteorology interesting, start watching the weather channel's weekly planner on TV and studying your newspaper's daily weather map when you start flight training. Develop a basic understanding of weather occurrences across the entire country. If you see a symbol that you don't understand, refer to the *Aviation Weather Services* manual.

When you understand that level of information, start studying the *Aviation Weather* manual and learn why those weather events occur. When you understand why something occurs, you learn with greater efficiency and far more enjoyment.

You must study the *Aviation Weather Services* manual.

Your instructor will show you several sources for obtaining official aviation weather information for flight planning: Telephone contact with FAA automated flight service stations, and internet contact with one or more of the numerous aviation weather websites. To decipher this information, you must once again study the *Aviation Weather Services* manual.

How do you use all of that information prior to a flight? The following list contains the most pertinent steps to follow when weather may be a problem. Common sense will allow you to eliminate some steps when the weather is obviously favorable.

- Seven days out—Start monitoring the weather channel's weekly planner.

- 48 hours out—Use the internet to monitor the various significant weather prognostic charts and the day-two convective outlook chart or the day-two convective outlook forecast (AC).

- 24 hours out—Use the internet to monitor terminal forecasts (TAFs) and the day-one convective outlook chart or the day-one convective outlook forecast (AC).

- 18 hours out—Use the internet to monitor the TAFs, the area forecasts (FAs), and the surface aviation weather observations (METARs).

An incorrect weather forecast is called a busted forecast.

Weather forecasters use a wide variety of resources, but due to unanticipated events, a forecast can become invalid. Pilots call this a *busted forecast;* consequently, your first weather-planning task is to validate the current forecast.

Compare the current TAFs and FAs to the current METARs. If the forecasts are correct or if the actual weather is better than forecast, you can treat the forecasts with a high level of confidence. For a long cross-country flight, you may be required to evaluate two or more area forecasts and several terminal forecasts. If you find a busted forecast, investigation is need. Why did this occur, and is an amended forecast available?

The inflight aviation weather advisories will reveal marginal or hazardous conditions.

If the forecasts looks good for your flight, your next objective is to check for hazardous weather. Those reports are called inflight aviation weather advisories—convective sigmets, sigmets, airmets, and urgent pilot reports (UUAs) as explained in the *Aviation Weather Services* manual.

- 12 hours out— Call flight service and request an outlook weather briefing.

- 6 hours out—Use the internet to obtain the current winds and temperatures aloft forecast (FD), and then complete your navigation log. Call flight service and request a standard weather briefing. You can file your flight plan at this time or wait until the next step.

- 1 to 3 hours out—Call flight service and request an abbreviated weather briefing for a few items such as winds, pilot reports (UAs), and your en route weather (state the time that you received the standard briefing). Use the internet to monitor METARs and the radar summary, surface analysis, and weather depiction charts. File your flight plan if you have not already done so.

Always call flight service just prior to departure for new TFRs.

- Just prior to departure—Call flight service to see if any temporary flight restrictions (TFRs) have been activated since your last weather briefing or if there have been any late-breaking weather changes. If you call flight service, the briefer will record your airplane's registration number, and this should protect you in the event a TFR is suddenly activated in the immediate area before you were able to contact ATC or the en route flight advisory service (EFAS, whose radio call sign is *flight watch*) when airborne.

Good pilots are weather wise. You started this process by watching the weather channel and reading your newspaper's daily weather map. Don't stop that activity. If you stay in tune with all the daily predictions, the aviation weather information that you receive from official sources when flight planning will make more sense.

FUEL

When you depart on a flight, federal regulations require you to have sufficient en route and reserve fuel for your flight based on using normal cruise power and the official information (weather, notams, TFRs) that was available when you departed. Three fuel categories exist:

If conditions are marginal, your en route fuel should always include a suitable diversion airport.

- En route fuel is that required for your planned flight. Your plan should always include a suitable diversion airport if there is any chance that weather conditions could deteriorate and become unflyable while en route to your destination.

- Reserve fuel must be at least 30 minutes during the day and 45 minutes at night.

- Contingency fuel is the remaining, usable fuel. You can use it for a deviation that you did not plan on in order to maintain the reserve fuel requirement.

Never assume that published fuel capacity or consumption rates are accurate.

Do not use an airplane's published fuel capacity or fuel consumption for flight planning if you are not intimately familiar with that airplane. Be conservative. A high-time engine or one with an inaccurate tachometer might burn more fuel; a deformed tank or the airplane's attitude when sitting on the ramp might reduce fuel capacity.

If the published fuel consumption at cruise power is 8.2 gallons per hour, add about ten percent and use 9 gallons per hour for flight planning purposes. If the performance chart indicates that 9 gallons per hour will give you a five-hour range, subtract 30 minutes and plan a 4-hour and 30-minute range for en route, reserve, and contingency fuel.

If you fly the same airplane frequently, you can confirm tank capacity and fuel consumption. Run a tank dry or drain it of the remaining fuel after a flight, and then refill it when the airplane is parked on level ground in its normal attitude. You now know the tank's capacity. Then, make certain that the fuel tanks are completely full before a flight, and note how much fuel is required to refill them after that flight. Use flight time and gallons added to compute the actual fuel consumption.

INFLIGHT PROCEDURES

Primary Considerations

- Looking outside the cockpit for conflicting traffic.

- Looking outside the cockpit in order to maintain attitude and heading. That's correct—attitude!

Your primary reference for maintaining heading is a distant point on the horizon.

The airplane's gyro-stabilized heading indicator is not your primary reference for maintaining heading. Yes, if you have set the heading indicator to agree with the magnetic compass, you use the heading indicator to turn to a new heading and you glance at it periodically to confirm that you are maintaining that heading.

Your primary reference for maintaining heading is a distant point on the horizon that appears to lie on your extended ground track. If you keep that spot in the same relative position on the airplane's windshield, you will maintain heading while looking outside the cockpit. Periodically, you glance at the heading indicator to confirm the heading, and when necessary, you pick a new reference point on the horizon.

- Looking outside the cockpit in order to evaluate surface wind conditions and possible forced landing sites.

- Looking outside the cockpit in order to locate the reference points that you circled on your navigation chart.

Time is the secret for finding the reference points that you circled on your navigation chart.

I mentioned previously, that the secret for finding these reference points is to use time as your primary reference. You mentally calculate that time by using the estimated groundspeed and distance. You never use a calculator and plotter to do this; that degree of accuracy is required only when you are flight planning or updating your navigation log during the flight.

Time Most humans can easily relate to time, whereas distances are somewhat ambiguous. I'll bet that if you're in your car traveling at 60 mph on an expressway and a road sign indicates that your turnoff is one mile ahead, you immediately realize that you have one minute to move over to the right lane, because 60 mph equals one mile per minute.

Use the same concept for finding your reference points. A poorly trained pilot will pass a reference point and then start looking in earnest for the next one, which may be ten miles ahead. He continually looks about but becomes confused due to surface features that are either misleading or contradictory to what he assumes when referencing his navigation chart. A properly training pilot will note the time at each reference point and estimate the en route time to the next point.

Navigation becomes easy when you base your actions on time.

Assume that your en route time estimate to an antenna-farm is five minutes. For the next four minutes, don't worry about or start looking for that reference. Maintain your heading and watch for traffic, surface wind indicators, and forced landing sites. When four minutes have elapsed, look ahead to where where the antenna farm should be located. Before you can say, "It should be right there," you will see it.

When you use time in this manner, it's best to use a clock's second hand or an electronic timer's seconds display. For example, if you pass over a reference point and it's 12 seconds past the minute. Just tell yourself that time started at a quarter past the minute and keep track of the minutes as they tick by. High accuracy is not required for this task.

Estimates based on simple rules-of-thumb work wonders.

Groundspeed You can quickly estimate the approximate flight time between reference points by using one of several groundspeed conversion techniques that are explained in the Mental Navigation Rules section, page 6 -18, of this chapter. The miles per minute technique and the equivalent fractions technique are the most common.

Let your fingers do the measuring.

Distance There are several ways to quickly measure the approximate distance between reference points on the chart. I use the middle segment of my little finger, but the finger used depends on hand size. Place your elbow on a table, hold your arm and extended fingers straight up, and then bend your fingers into an inverted-U—each finger tip will touch its base. Now, compare the full width of each finger's inverted-U to the navigation chart's mileage scale. Use whichever finger measures 10 miles on a sectional chart. That distance then represents 5 miles on a terminal area chart and 20 miles on a world aeronautical (WAC) chart.

You can also extend your middle and index fingers together, and place them on the chart mileage scale. If the full width of both fingers gives you the 5, 10, and 20 mileage measurements, you can use that technique.

Each tic mark on a sectional chart's meridians equals one nautical mile.

A third method requires a pencil or some other linear object. (When working with these reference points you are determining distances that are seldom more than 10 miles apart.) Place the tip of the pencil on one reference point and use your fingers to hold the pencil at the second reference point. Now, move the pencil to a mileage scale or a meridian (line of longitude) on the navigation chart to determine the mileage. For sectional charts, each degree of latitude equals 60 nautical miles, and therefore, each tic mark on a meridian equals one nautical mile. With the exception of the equator, which is also a great circle like the meridians, you cannot use the tic marks on the parallels (lines of latitude), because those distances decrease as latitude increases.

In summary, the technique that I have described for determining time can be accomplished in three or four seconds. Approaching a reference point, you estimate the distance to the next reference point, and by using a groundspeed conversion technique, you convert that distance to time. That's it!

When you navigate in this manner, you minimize your workload, gain extra time for cockpit management, and constantly remain aware of where you've been, where you are, and where you're going. Today, that's called situational awareness; a mandatory requirement for competent airmanship.

- If you fail to find a circled reference point, do not continue. Return to the last reference point and try again. As someone once said, "You aren't lost when you realize it, you've been lost ever since you passed your last known reference point."

Secondary Considerations

Confirm the winds aloft as soon as possible.

- Compute actual groundspeed in order to validate your planned groundspeed, fuel consumption, and estimated arrival times at subsequent checkpoints and your destination.

- Update your en route and destination weather and obtain pilot reports (pireps) by contacting en route flight advisory service (Flight Watch on 122.0 mHz). Normally, you make this call at five minutes or so past the hour so that you get the latest weather reports. If weather conditions are a concern, you should make this call at least every hour. If you are unable to contact Flight Watch—always your first choice—contact a flight service station (FSS) to obtain the same information.

Use Flight Watch for weather and pireps.

- Whenever you are in the vicinity of a VOR that provides hazardous inflight weather advisory service (HIWAS), listen to see if new advisories have been issued—airmets, sigmets, convective sigmets, or urgent pilot reports.

Use HIWAS for late developing flight hazards.

- Utilize ATC's radar advisory service for traffic advisories. An Aircraft Owner's and Pilot's Association (AOPA) safety advisory publication stated that eighty-two percent of midair collisions occur when a faster airplane overtakes a slower airplane. You can't see what's behind you, but ATC can when you fly at an altitude that has radar coverage. They will also inform you if a temporary flight restriction (TFR) is suddenly activated on your route of flight.

Use radar advisory service for traffic advisories.

The AOPA (www.aopa.org) is an excellent organization that provides numerous benefits and resources for all pilots. Aircraft ownership is not required, and the membership dues are very reasonable considering what you receive in return.

Additional Operating Procedures

- File a flight plan for all extended cross-country flights so that search and rescue service is notified if you do not complete your flight. If you change your route during the flight, advise a FSS of that change. When flying in remote areas, make a VFR position report to a FSS every thirty minutes if possible.

- Never depart an airport unless you have checked the government's Airport/Facility Directory (A/FD) for special departure procedures and airport diagrams. This official publication is revised every 56 days. You must have an airport diagram in your possession whenever you operate at airports that have unique or complex runway and taxiway layouts.

- Always note your takeoff time, and record it on your navigation log when time permits.

• All departures require a climb to pattern altitude if you are departing the area. If your departure heading takes you back toward the traffic pattern, do not turn more than 45 degrees until you are well clear of aircraft that may be inbound on the traffic pattern's entry leg. If you're remaining in the traffic pattern or making a downwind departure, you can turn onto the crosswind leg when you're 300 feet below pattern altitude.

• If you are flying to an unfamiliar, controlled airport using basic navigation procedures and you have not been using the techniques described in this chapter, never call ATC for landing clearance until you locate that airport visually.

If you do not use the techniques that I have described when flying to a controlled airport using basic navigation procedures, do not call ATC for landing clearance until you locate that airport visually.

You're asking for trouble if you call ATC when you're not absolutely sure of your position and the airport's location. A minute or two after the call, ATC will usually ask for your position, and that's when the confusion and pressure really increase. Stay above the Class-D airspace until you find the airport and then call overhead for landing clearance.

The following illustration indicates the ideal flight path after you report overhead the field and are cleared to enter the traffic pattern. There are many ways to do this, but this procedure works exceptionally well when visibility is marginal.

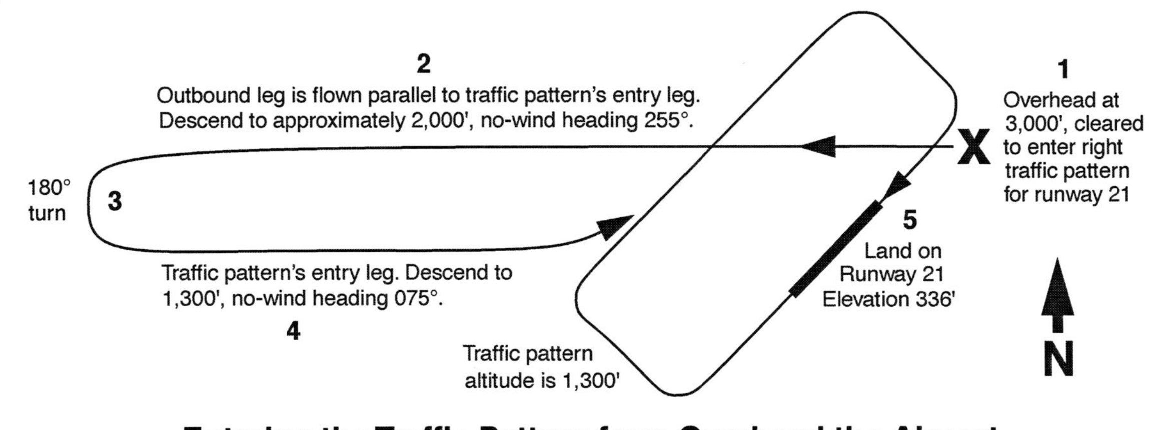

Entering the Traffic Pattern from Overhead the Airport

• While entering the traffic pattern at an unfamiliar airport, plan ahead and study the surrounding area for emergency landing sites in the event your engine fails during departure.

- A standard traffic pattern entry requires a 45-degree entry leg prior to the downwind leg. This entry leg should aim at the center of the landing runway in order to avoid conflicts with departing traffic.

LOST PROCEDURES

If you are lost, follow the four-C's rule.

When I hear stories about pilots who were lost, I immediately question their training and their personal priorities. Too many instructors place the primary emphasis on radio navigation instead of the basic, visual navigation procedures that are explained in this chapter, and too many pilots who were taught the proper skills fail to maintain them. A pilot who is lost, however, should always follow the four-C's rule: *climb*, *communicate* (with ATC), *confess* (your situation), and *comply* (with ATC instructions).

Lost communications should start on the standard FSS frequency, 122.2 mHz. Other frequencies are listed on navigation charts at VORs and remote communication outlets (RCOs), but if you are lost, you would not know which of those to use. If you do make contact with FSS, they can provide you with ATC frequencies, and in some cases they can determine your position by using direction finding equipment.

When you run out of ideas, use 121.5 mHz for help. An emergency situation is not required.

If you are unable to contact FSS, use the emergency frequency, 121.5 mHz. Your initial transmission should be, "Any station, Skyroamer 9117R, transmitting 121.5." More than one facility will probably answer your call, so pick the one you want to talk to and say, "All stations standby, Denver center, Skyroamer 9117R is lost and needs guidance to an airport that has aviation fuel".

In the next chapter, you will learn how to determine your position using radio navigation, providing you are high enough to receive such facilities. After you determine your position, you must use the following diversion procedure in order to fly to a suitable airport.

DIVERSION PROCEDURE

When an inflight diversion is required—a change of route or destination—you must know how to determine the new course, new heading, distance, flight time and fuel consumption without using a plotter or calculator.

If lost, put the airplane in a shallow-bank turn and circle while you determine your position.

- Put the airplane into a shallow-bank turn so that you remain over your known position while you solve this problem. You need both hands to work with your navigation chart, so trim the elevator for hands-off flight and use your feet on the rudder pedals to maintain the bank attitude. If you let the bank shallow out, you will gain altitude; if you let the bank steepen, you will lose altitude. While you work the diversion problem, continue to scan for traffic and monitor bank angle.

- Depict the course on your chart by using one of two methods. Using a straight edge such as the edge of a checklist, draw a line from your present position to your diversion airport. Or, using the thumb and index finger of each hand, pinch the chart at the leg's end points, pull your hands apart to stretch and fold the chart, and rub this fold back and forth across your chin to form a sharp crease.

- Determine the magnetic course by comparing that course line to a compass rose at a nearby VOR station. Use any straight edge—a pencil or the edge of a checklist—to do this. Align the straight edge with the course and keep it parallel to that course while you slide it over to the center of the compass rose. Read the magnetic course on the compass rose, but be careful to consider your intended direction of flight and to interpret the course numbers correctly. Misinterpretation is a common error.

- Determine the magnetic heading by applying a reasonable wind correction angle to the magnetic course. The mental navigation rules in the next section explain how this is done.

Use rules of thumb to determine the heading and en route time to the diversion airport.

- Determine the distance to destination by using the previously described finger or pencil method.

- Determine the time to destination by using the previously described groundspeed conversion techniques.

- Determine the fuel consumption for this leg.

MENTAL NAVIGATION RULES

Reciprocals

The Double-2 rule. The reciprocal of 147 is 327. To make that conversion, the first number of the reciprocal must be a 0, 1, 2, or 3. Do that by adding or subtracting a 2 to the first digit of the original number (147). To the second digit of the original number, use 2 again, but perform the opposite function. If 2 was added to the first digit, subtract 2 from the second digit and vice versa. The last digit always remains the same. Examples: 246 becomes 066, 183 becomes 363 or 003, 341 becomes 161, and 039 becomes 219.

Distance To Flight Time Conversions

When you can quickly convert distance into flight time, estimated arrival times at reference points are easy to determine. Flight management is simplified when you think in terms of time rather than distance. Learn all three conversion methods so that the most convenient one can be utilized.

Method I: Equivalent Fractions Change the fraction that represents groundspeed $\left(\frac{\text{Groundspeed}}{60}\right)$ to an equivalent fraction whose numerator equals the distance to be flown. The denominator will then equal the flight time.

Moving decimal points Groundspeed equals 140 knots, and the decimal point is moved in both the numerator and the denominator to obtain equivalent fractions $\left(\frac{140}{60} = \frac{14}{6} = \frac{1.4}{0.6}\right)$. Using these fractions, 140 nautical miles (nm) would take 60 minutes, 14 nm would take 6 minutes, and 1.4 nm would take 0.6 minutes. For other mileages, find a combination of those fractions whose numerators add up to the desired mileage. Add the denominators of those fractions to obtain flight time. 17 nm would take 7 minutes: $\left(\frac{14}{6,} \frac{1.4}{0.6,} \frac{1.4}{0.6,} \frac{\text{added} =}{\text{added} =} \frac{17}{7}\right)$

Dividing Groundspeed equals 160 knots, and both the numerator and the denominator are divided by 2 to obtain equivalent fractions $\left(\frac{160}{60} = \frac{80}{30} = \frac{40}{15} = \frac{20}{7.5}\right)$. For other mileages, find a combination of those fractions whose numerators add up to the desired mileage. Add the denominators of those fractions to obtain flight time. 60 nm would take 22 minutes: $\left(\frac{40}{15,} \frac{20}{7.5,} \frac{\text{added} =}{\text{added} =} \frac{60}{22}\right)$

Method 2: Even Miles per Minute A groundspeed of 60, 120, or 180 knots is equivalent to 1, 2, and 3 miles-per-minute, respectively. Time equals mileage divided by miles-per-minute. If groundspeed is 120 knots and distance is 18 miles, time for that leg is 9 minutes—18 divided by 2.

Method 3: Adjusted Miles per Minute This method carries the miles per minute method one step further and is accurate for distances up to about 80 nautical miles. Compute the time using 1, 2, or 3 miles per minute, and then, using the following table, add or subtract a percentage of that time based on actual groundspeed. This method is used by pilots who always fly the same airplane since only one relationship in the following table must be remembered.

100 kts = 120 kts + 20%	2 and 3 nm per min.
110 kts = 120 kts + 10%	160 kts = 180 kts + 20%
120 kts = 2 nm per min.	170 kts = 180 kts + 10%
130 kts = 120 kts - 10%	180 kts = 3 nm per min.
140 kts = 120 kts - 20%	190 kts = 180 kts - 10%
150 kts = the average of	200 kts = 180 kts - 20%

Example: Groundspeed is 140 knots; distance is 12 miles. At 120 knots or 2 miles per minute, 12 miles would take 6 minutes. Subtracting 20% of 6, or 1 minute, yields 5 minutes.

Flight Time to Groundspeed Conversion

Determine groundspeed by using equivalent fractions but reversing the distance to flight time procedure. Form a fraction by placing distance over the time required to fly that distance. Now, pick a number that when multiplied by the denominator yields approximately 60. Groundspeed equals that number multiplied by the fraction's numerator.

Examples: 12 minutes is required to fly 27 nm ($\frac{27}{12}$). 12 times 5 equals 60; therefore, groundspeed equals 5 times 27 or 135 knots. 10 minutes is required to fly 25 nm ($\frac{25}{10}$), 10 times 6 equals 60; therefore, groundspeed equals 6 times 25 or 150 knots. 8 minutes is required to fly 21 miles ($\frac{21}{8}$), 8 times 8 equals 64; therefore, groundspeed is slightly less than 8 times 21 or 168 knots. 160 knots is a good guesstimate.

True Airspeed

To determine true airspeed (TAS), increase indicated airspeed (IAS) by 2 percent per thousand feet of altitude.

Example: IAS is 110 knots at 7,000 feet. For TAS add 14 percent (2 times 7) to the IAS. To add 14 percent, add 10 percent (easy to compute) and then 4 percent. 10 percent of 110 is 11, and 4 percent is approximately 4. The 14 percent addition equals 15 knots (11 plus 4) which gives an approximate TAS of 125 knots (110 plus 15).

Example: IAS is 135 knots at 9,000 feet. For TAS add 18 percent to the IAS. 10 percent of 135 is 14 knots, and 8 percent is about 10. The 18 percent increase equals 24 knots, which gives an approximate TAS of 160 knots.

Wind Components

Wind can be quickly resolved into crosswind and headwind or tailwind components. The required information is wind velocity, wind direction relative to the runway or the airplane's nose or tail, and the easy-to-memorize chart on the next page.

To help you memorize the chart, note that when the wind is at the 45-degree position, the component multiplier is always 0.7. As you move in 15-degree increments away from the 45-degree position, adjust the 0.7 multiplier by 0.2. If the wind formed a 60-degree angle (45 plus 15) with the airplane's nose, the crosswind multiplier would be 0.9 (0.7 plus 0.2) and the headwind multiplier would be 0.5 (0.7 minus 0.2).

Example: Heading is 090 degrees, and the wind is 165 degrees at 14 knots. Since the wind is 75 degrees off the nose, the headwind multiplier is 0.3, so the headwind component is 4 knots (14 times 0.3). The crosswind multiplier is 1.0, so the crosswind component is 14 knots.

Example: Heading is 360 degrees, and the wind is 150 degrees at 24 knots. Since the wind is 30 degrees off the tail, the tailwind multiplier is 0.9, so the tailwind component is 22 knots. The crosswind multiplier is 0.5, so the crosswind component is 12 knots.

Remember, in the northern hemisphere, the wind shifts in a counterclockwise direction as the airplane descends towards the surface. The wind shifts in a clockwise direction as the airplane leaves the surface or passes through a weather front.

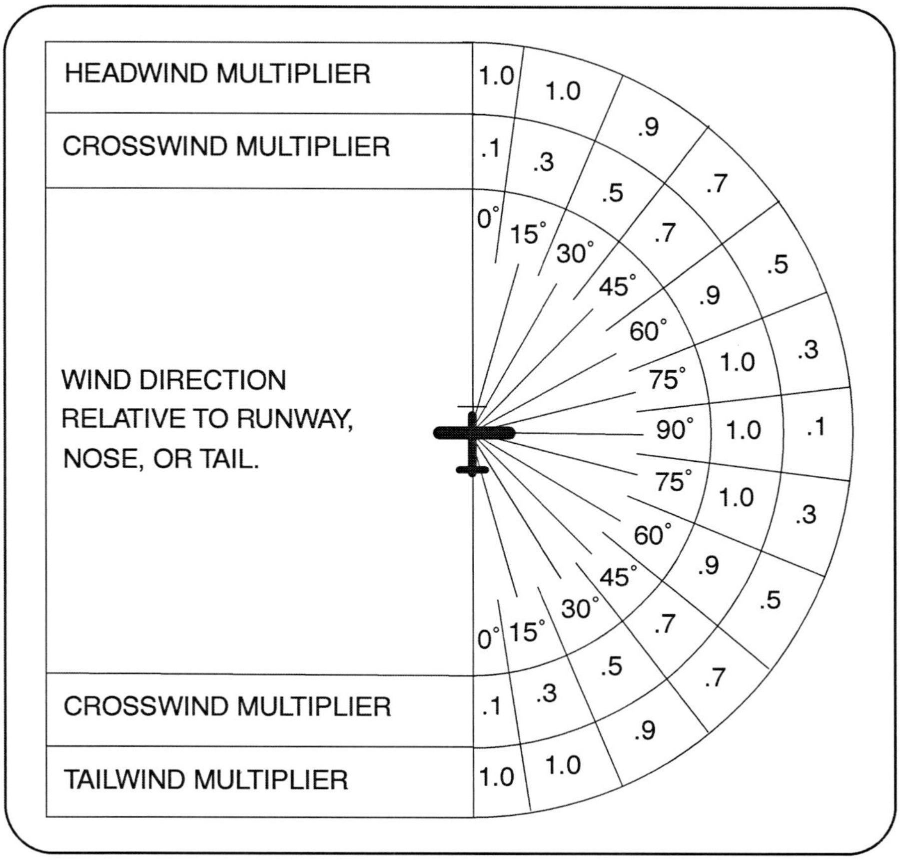

Determining Wind Correction Angles

Have you ever seen an experienced pilot takeoff, intercept a course, and roll out on a heading that results in a perfect track? That professional trick is easy to do if you know the rules of thumb for determining wind components and wind correction angles.

To determine the wind correction angle (WCA), you must know the crosswind component. Then, divide the crosswind component by the TAS (in miles per minute) for the answer.

Example: The crosswind component is 17 knots and TAS is 2 miles per minute (120 knots). The WCA is 8.5 degrees (17 divided by 2).

Example: The crosswind component is 12 knots and TAS is 3 miles per minute (180 knots). The WCA is 4 degrees (12 divided by 3).

Example: The crosswind component is 15 knots, and TAS is 140 knots, somewhat faster than 2 miles per minute. At 2 miles per minute the WCA would be 7.5 degrees; at 3 miles per minute it would be 5 degrees, so 6 or 7 degrees would be a good guesstimate.

For visual navigation as explained in this chapter, you can simplify the WCA process even further. The only requirement is that you know the wind speed at your altitude. Now, if it is a direct crosswind, divide it by your approximate TAS in miles per minute to determine the WCA. If the wind is 45 degrees off your nose or tail, multiply the wind velocity by 0.7 to determine the crosswind component, and divide it by your approximate TAS in miles per minute to determine the WCA. If the wind is from any other direction, make a guesstimate of the WCA based on what it would be for the direct or the 45-degree crosswind.

Density Altitude, Approximate

Field elevation plus or minus 120 times the difference between the actual and standard Celsius temperatures.

Example: Field elevation is 3,500 feet, temperature is 110° F, 43° C. Standard temperature in °C is always 15 minus twice the altitude in thousands of feet; in this case 8° C, 15 minus 7. Approximate density altitude: 3,500+120(43-8) = 7,700 feet.

Adding Minutes on a Calculator

If your log book is kept in hours and minutes, here is an easy way to add those times on a calculator. Add two zeros after the hour of each entry and then add all times. Finally, continuously add 9940 until the minutes decrease below 60.

Example: Add 1:55, 1:09, and 0:59. The calculator inputs would be 10055 plus 10009 plus 59 which gives an answer of 20123. Now, add 9940 until the minutes decrease below 60. The first 9940 yields 30063, and the second 9940 yields 40003. To obtain the answer, 4:03, remove the two zeros between the hours, 4, and the minutes, 03.

REVIEW TEST

1. What are the three elements of aviation safety?
2. What are the three basic navigation procedures?
3. Where should your route start and end when planning a dead reckoning route?
4. How often should you circle reference points along a dead reckoning route?
5. After you level off at cruise altitude, what should you check as soon as possible?
6. What navigation term is not listed on most navigation logs and why is it important?
7. What is the rule for remembering how to apply variation to the true course?
8. What is the sequence for determining compass heading from true course and the sequence for determining magnetic course from true course?
9. How can you monitor the forecast weather starting seven days before a flight?
10. Before you start planning a flight in earnest, you monitor the weather prognostic charts, the weather forecasts, and actual weather. Why is this important?
11. Just before your departure, should you check with flight service?
12. How do you treat published fuel consumption and fuel capacity when flying an airplane that you are not intimately familiar with?
13. What are your primary considerations during a cross-country flight?
14. What primary reference do you use so that you can find your reference points during a cross-country flight?
15. What is a quick method for determining short distances during a cross-country flight?
16. What are your secondary considerations during a cross-country flight?
17. Before you depart an airport what must you determine by regulation?
18. If operating from or to a complex airport, what reference must you have?
19. What should you do if you cannot find a circled reference point on your chart?
20. If lost, what rule should you follow?
21. You must divert to an unplanned airport. Do you need a plotter and flight calculator to do this?
22. Mentally compute the reciprocal of 138 using the 2 ± 2 rule.
23. Groundspeed is 110 knots. Mentally compute how long it will take to fly 37 miles.
24. It took 42 minutes to fly 67 miles. Mentally compute your approximate groundspeed.
25. At 5,500 feet, your indicated airspeed is 105 knots. Mentally compute the true airspeed.
26. You have a 40 knot quartering headwind that is 30 degrees off the nose. Mentally compute the headwind and crosswind components.
27. Your groundspeed is approximately 120 knots or 2 miles per minute, and your next reference point is 7 miles ahead. What is your approximate en route time?

SITUATIONAL AWARNESS!

A MODERN TERM THAT HAS

SUPERSEDED AN OLD ADAGE:

WHERE HAVE YOU BEEN?

WHERE ARE YOU NOW?

WHERE ARE YOU GOING?

CHAPTER 7

RADIO NAVIGATION: ORIENTATION AND TRACKING

Radio navigation uses very high frequency, omni-directional ranges (VORs) and low or medium frequency nondirectional beacons (NDBs). These systems will eventually be replaced by the global positioning system (GPS)—discussed in Chapter 8. Until that happens, you must master radio navigation. Orientation: You determine your position in relation to a VOR or NDB station or course. Tracking: You correct for wind drift in order to remain on a VOR or NDB course.

VOR ORIENTATION

GENERAL

A VOR station emits a continuous, variable-phase signal in all directions. The airplane's VOR receiver measures the signal's phase at your location. To determine your magnetic bearing to or from the VOR, you must rotate the omnibearing selector's (OBS) course knob until the course deviation indicator (CDI) is centered with a To or a From indication.

The following illustration depicts an OBS and the 360 radials that surround a VOR station.

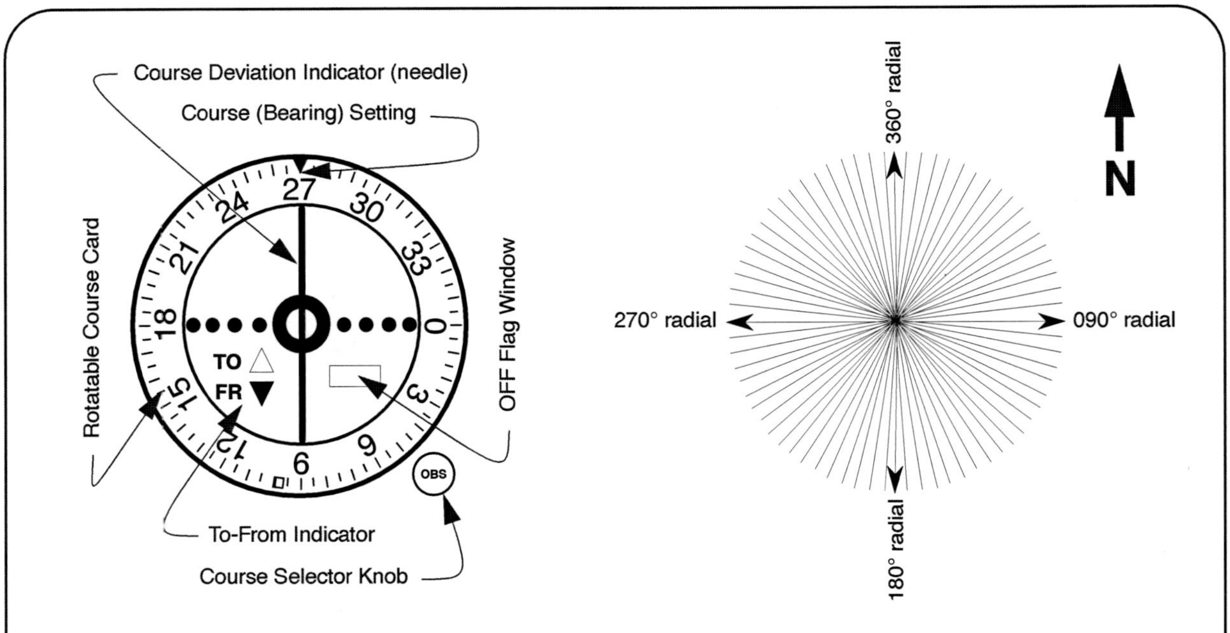

Omni-Bearing Selector (OBS) **VOR Radials**

VOR CONCEPTS

If you flew continuous steep turns while tuned to a distant VOR station with the CDI centered, the CDI and the To-From indicator would not move.

The CDI does not move because the steep turn keeps you on the same VOR radial. The To-From flag does not move because the turn keeps you in the same geographical position relative to the VOR station.

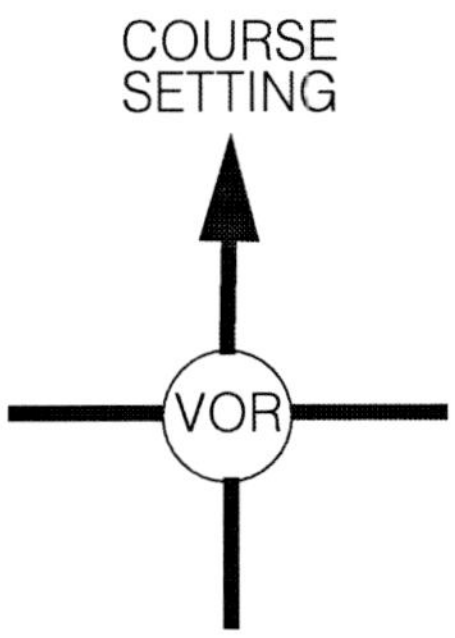

The VOR Tuning Cross

The VOR tuning cross is an imaginary cross that sits atop all VOR stations. When you rotate the OBS's course card, you are rotating the tuning cross so that the arrowhead leg points in the direction that is designated by the course setting at the top of the OBS's course card.

VOR receiver operation depends on two factors: Your location in relation to the VOR, and the orientation of the tuning cross. To understand these relationships, refer to the expanded VOR tuning cross illustration on page 7 - 5.

In the expanded illustration, the OBS course card is set to 360 degrees, denoted as zero on the card. 360 is the proper term since zero is never spoken when referring to courses, bearings, or radials. Like most of aviation's numerical references, the numerals are spoken individually—3-6-0.

The tuning cross's two, perpendicular legs divide the VOR's airspace into four quadrants. The course card is set to 360 degrees, so the vertical, arrowhead leg represents the 360-degree radial when north of the VOR and the 180-degree radial when south of the VOR. The second, horizontal leg represents the 90-degree radial when east of the VOR and the 270-degree radial when west of the VOR.

A VOR radial is a magnetic bearing from the VOR, but *course* is the term you use when tracking a radial to or from a VOR station. If tracking outbound on the 360-degree radial, your course is 360 degrees; if tracking inbound, it is 180 degrees.

The expanded illustration shows eight OBSs, two on each leg and one in each quadrant of the tuning cross. They represent the aircraft's position, A through H, in relation to the tuning cross and the VOR station. At each position, *regardless of the aircraft's heading*, the depicted CDI and To-From flag indications will always exist. Using this knowledge, examine each indication and verify the following true statements:

- The position of the CDI depends on the aircraft's location in relation to the arrowhead leg of the tuning cross, the course designated on the OBS's course card.

- The position of the To-From flag depends on the aircraft's location in relation to the second leg of the tuning cross, which is always perpendicular to the arrowhead leg.

- The CDI tells a lie—points away from the desired course—when the airplane's heading is more than 90 degrees from the OBS's course setting. This condition is called reverse sensing.

- The CDI tells the truth—points toward the desired course—when the airplane's heading is within 90 degrees of the OBS's course setting. This condition is called like-heading, like-bearing.

You must comply with the like-heading, like-bearing rule. Reverse sensing is avoided like the plague.

To avoid confusion during VOR navigation, you must comply with the like-heading, like-bearing rule—the airplane's heading and the OBS course setting must be within 90 degrees of each other so that the CDI points toward the course. If that is not the case, you will have reverse sensing.

VOR OPERATING RULES

Tuning the Receiver

Is the VOR receiver properly tuned? The abbreviation VOR has three letters, and this serves as a reminder that three items must be accomplished when you tune a station: tune the frequency, identify the station, and set the OBS.

To minimize workload, memorize the Morse Code.

To minimize workload when identifying navigation stations, memorize the Morse Code. It's easy to do if you just learn three new letters each day. Use a flashlight to flash the codes on a ceiling or wall, or make up a set of flash cards. This important skill is not required for private pilot certification.

Accuracy Checks

Is the VOR receiver operating properly? For instrument flying, federal regulations require a VOR accuracy check within the preceding 30 days by using a VOR test signal, a VOR ground or airborne reference point, or a dual comparison check—two VOR receivers checked against each other. Regardless of regulatory requirements, all pilots who use VOR equipment should check receiver accuracy before they use a VOR receiver for navigation.

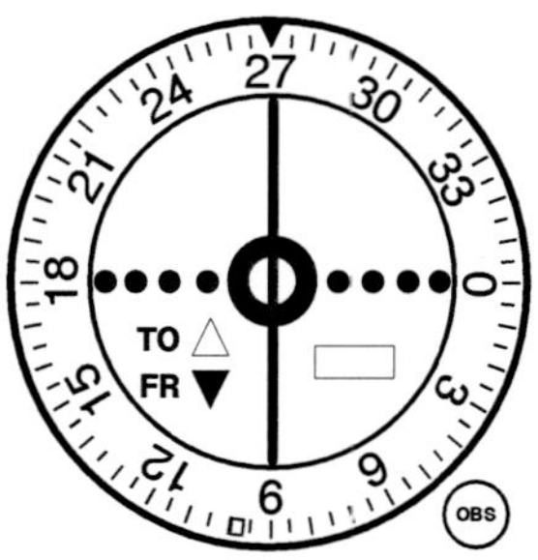

Station Orientation

VOR Station Orientation

Where is the airplane in relation to the VOR station? Center the CDI with a From indication by turning the OBS's course knob. The bearing from the station (radial) is now indicated under the course setting at the top of the OBS's course card.

The adjacent illustration shows a centered CDI, a From indication, and 270 degrees on the OBS course card. The airplane is west of the VOR on the 270-degree radial.

VOR Position Triangulation

To determine your exact position on a navigation chart, place the airplane in a continuous turn so that you remain over that

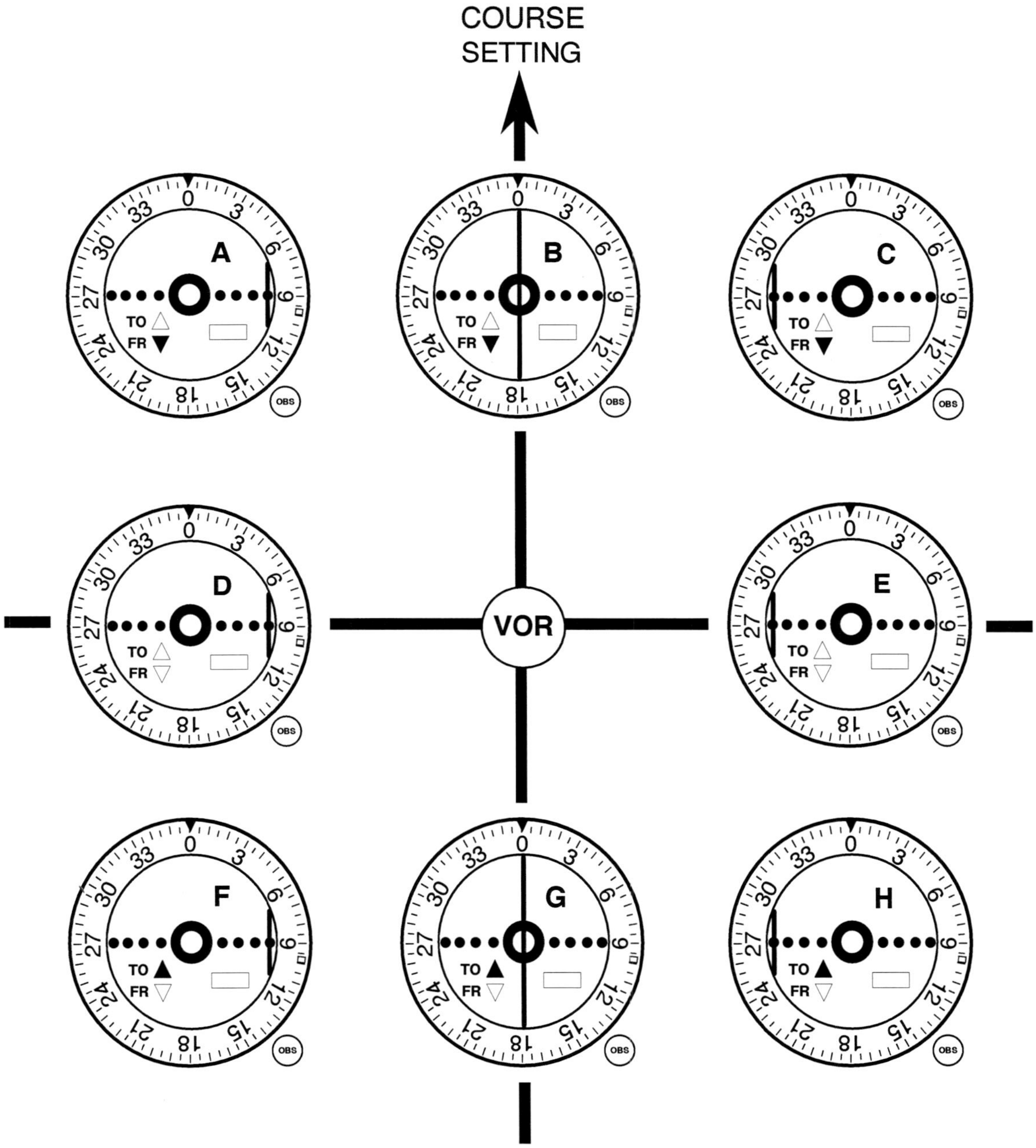

The Expanded VOR Tuning Cross

location, and using the VOR station orientation procedure, determine the radials you are on using two or three VORs. Plot those radials on your navigation chart. You are located at the point where the radials cross each other.

VOR Course Orientation

Where is the VOR course? Turn parallel to the course to ensure like-heading, like-bearing, and set the OBS to the desired course. The CDI will point toward that course.

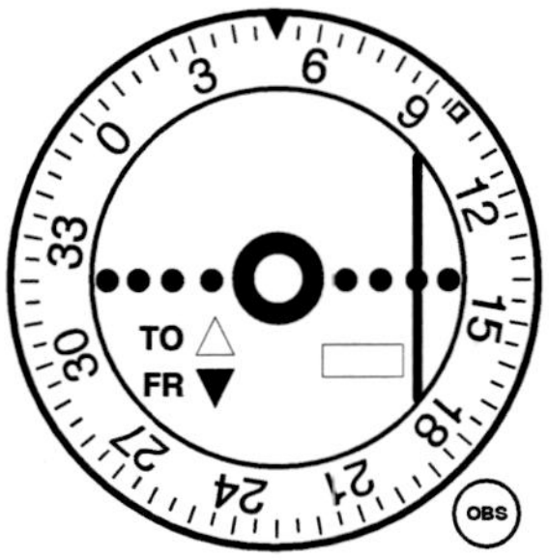

Course Orientation

In the adjacent illustration, the airplane was turned to 050 degrees in order to parallel the 050-degree course. The OBS was then set at 050 degrees, and because the CDI is to the right, the course is to the right.

VOR On-Course Indication

Is the airplane on the VOR course? Yes, providing the CDI is centered.

VOR Cross-Bearing Orientation

Where is a VOR cross-bearing—a VOR radial that crosses the current course? Set the cross-bearing (radial) under the course setting on the OBS's course card.

The OBS will now display a From indication and one of three CDI positions as depicted in the next illustration. If the CDI points toward the VOR station that generates the cross-bearing, that radial is ahead of you. If the CDI is centered, you are on that radial. If the CDI points away from the VOR station that forms the cross-bearing, that radial is behind you.

In the illustration, the airplane is flying from VOR-A to VOR-C. A cross-bearing, the 270-degree radial, is generated by VOR-B. The OBS is set at 270 degrees, and the CDI positions are shown prior to, at, and after crossing the 270-degree radial.

VOR OPERATING RULES SUMMARIZED

- Is the VOR receiver properly tuned? Frequency, identifier, OBS.

- Is the VOR receiver operating properly? Use one of the four methods to check it's accuracy.

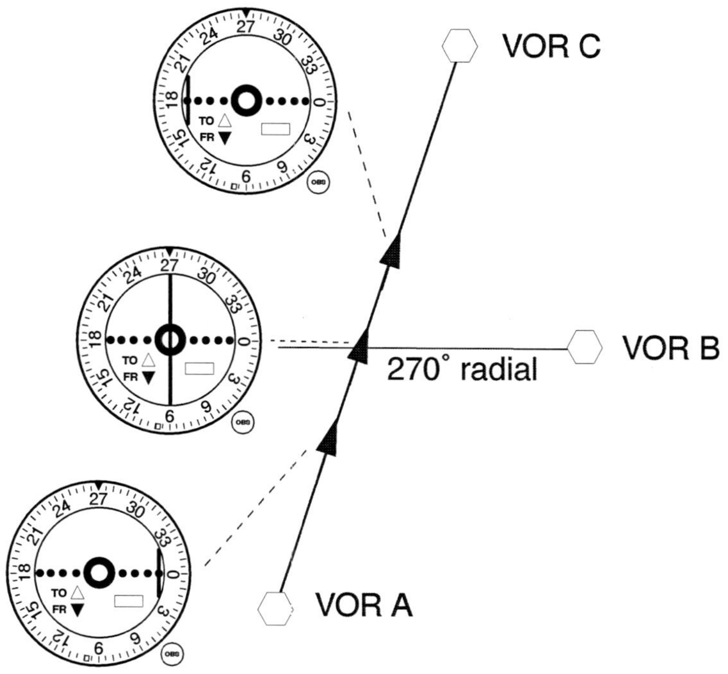

VOR Cross-Bearing Orientation

- Where is the airplane in relation to the VOR station? Center the CDI with a From indication. The OBS's course setting indicates the bearing from the station (radial).

- Where is the VOR course? Parallel the course, and set it under OBS's course setting. The CDI will now point to the course because you have like-heading, like-bearing.

- When is the airplane on the desired VOR course? When the CDI is centered.

- Where is the VOR cross-bearing? Set the cross-bearing (radial) on the OBS's course card with a From indication. If the CDI points toward the VOR that generates the cross-bearing, that radial is ahead of you. If the CDI points away from the VOR, that radial is behind you. If the CDI is centered, you are on that radial.

- If a VOR signal fails, an off-flag will appear on the OBS.

ADF ORIENTATION

GENERAL

A non-directional radio beacon (NDB) emits a continuous signal in all directions. The airplane's automatic direction finder (ADF) processes that signal and provides bearing information on the ADF radio compass as depicted on the left. The bearing pointer always points at the selected station, which can be either an NDB or an AM radio station. Both types of stations are depicted on aeronautical charts, and both provide useful information for radio navigation.

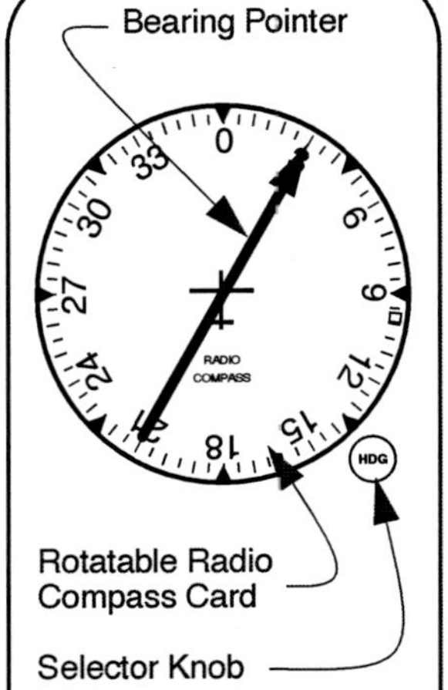

ADF Radio Compass

Triple check the ADF's mode selector—OFF, REC or ANT (receive or antenna), and ADF. When the radio is on, leave the mode selector in the ADF position. You will become disoriented if you follow a bearing pointer that was inoperative because the mode selector was in the REC or ANT position, a position that gives only audio reception. ADF receivers do not have an OFF flag, so you must monitor the station's audio identifier to confirm operation.

In the REC or ANT position, only the sense antenna is active for audio reception. In the ADF position, the sense and loop antennas are both active for audio and bearing information. In older airplanes, the sense antenna is a long wire located above the fuselage, and the loop antenna is located in a housing below the fuselage—both antennas are in the same housing on newer airplanes. Airplane's that have a factory-installed GPS receiver seldom have an ADF receiver.

ADF CONCEPTS

If you flew continuous steep turns while tuned to a distant NDB, the ADF bearing pointer *would not move*. It always points at the NDB. The perceived movement results from the airplane and radio compass card turning beneath the bearing pointer. Mastering ADF is easy if you think in this manner.

The steep turn keeps the airplane on the same bearing from the station (same as a VOR radial). Since the bearing from the NDB is not changing, the bearing pointer does not move—it always points to the station.

The operation of ADF equipment depends only on the aircraft's location in relation to the NDB. Unlike the movement of a VOR receiver's OBS course card, movement of the rotatable radio compass card does not affect the ADF bearing pointer.

Normally the radio compass card is set to the parked position with zero at the top.

You rotate the radio compass card only when you orient your position from a station. In all other situations, the card is left in the *parked position* with zero at the top.

The information in the next three paragraphs and the ADF tracking information in the next section are not required for private pilot certification. This information is presented to help those who are interested in mastering ADF tracking.

To simplify ADF flying with a parked compass card, I use the term deviation angle as depicted in the adjacent illustrations. Aircraft movement causes the bearing pointer to deviate. Depending on the current task, bearing-pointer deviation may or may not be desired.

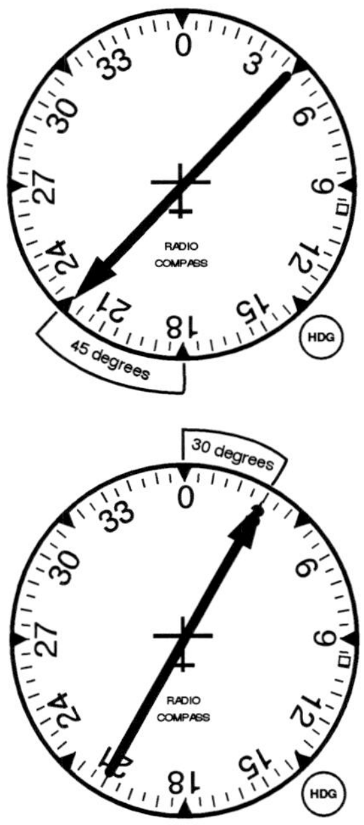

Deviation Angles

Whenever the radio compass card is parked, deviation angles are used, and the numbers on the face of the card are ignored. Whenever the radio compass card is rotated for station orientation, the radio compass card numbers are used, and the deviation angle is ignored.

The top radio compass in the adjacent illustration indicates an NDB that is behind the airplane. The deviation angle is the 45-degree angle between the head of the bearing pointer and the bottom of the radio compass card—the airplane's tail. The lower radio compass indicates an NDB that is ahead of the airplane. The deviation angle is the 30-degree angle between the head of the bearing pointer and the top of the radio compass card—the airplane's nose.

ADF OPERATING RULES

Tuning the Receiver

Is the ADF receiver properly tuned? The abbreviation ADF has three letters, and just like the VOR abbreviation, this serves as a reminder that three items must be accomplished when tuning a station: tune the frequency, identify the station, and set the radio compass card.

Accuracy and Sensitivity Checks

Is the ADF receiver operating properly? When an NDB or commercial broadcast station is tuned, the ADF bearing pointer must point at and lock onto that station. If the bearing pointer wanders about, a problem exists.

You can check the receiver's sensitivity if it has a test switch. When the switch is pushed, the bearing pointer will be driven away from its tuned position. When the switch is released, the pointer should snap back to the tuned position. If the pointer's return to the tuned position is sluggish, a problem exists. If the receiver does not have a test switch, you can conduct the same test by momentarily tuning a second station and then retuning the first station.

ADF Station Orientation

Where is the airplane in relation to an NDB or an AM radio station? Tune and identify the station, and set the airplane's magnetic heading at the top of the radio compass card. The tail of the ADF bearing pointer now indicates the bearing from that station (same as a VOR radial).

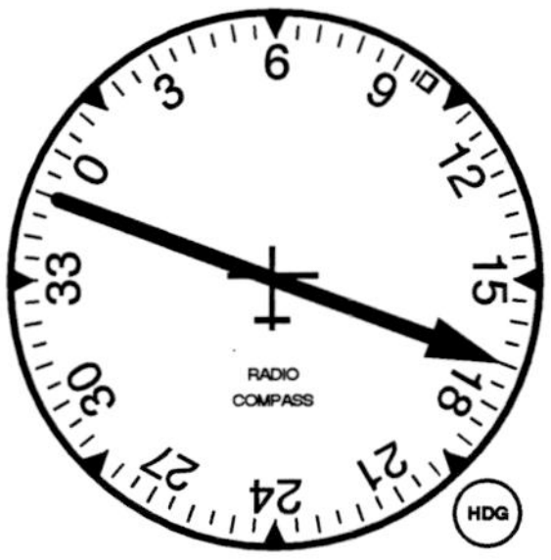

Station Orientation

The adjacent illustration shows a compass card that has been rotated to the airplane's heading, 060 degrees. The tail of the ADF bearing pointer indicates the bearing from the station, 350 degrees. The airplane is north of the NDB on the 350-degree bearing from the station.

ADF compass card rotation rule

- Rotate the card to the airplane's magnetic heading for NDB or AM radio station orientation.

- Park the card at zero when tracking to or from an NDB or a AM radio station.

ADF OPERATING RULES SUMMARIZED

- Is the ADF receiver properly tuned? Frequency, identifier, and radio compass card setting.

- Is the ADF receiver operating properly? Points to station, does not wander about, and snaps back to the tuned position when it is intentionally deflected.

• Where is the airplane in relation to the NDB station? Set your magnetic heading at the top of the radio compass card; the tail of the pointer indicates your bearing from the station.

TRACKING

How do you stay on course? By tracking, a procedure that uses specific heading changes to quickly establish the proper wind correction angle and keep the airplane on course.

The approximate wind correction angle can be determined after two to four heading changes.

Tracking is another application of the two-step correction rule: Make a correction that will obviously work, and when that occurs, remove half of the original input. With this procedure, the wind correction angle is determined after only two to four heading changes.

VOR TRACKING

To track a VOR course, turn to the no-wind heading when you intercept the course. The CDI will now be centered. If drift occurs, turn 20 degrees into the wind. When back on course, remove half of the 20-degree correction and fly with a 10-degree wind correction angle. One of three things will then happen: The airplane will either remain on course or it will drift downwind or upwind of the course.

Airplane remains on course The CDI remains centered.

Airplane drifts downwind of course Reestablish the original 20-degree correction as depicted in the illustration on page 7-13. When back on course, turn 5 degrees toward the course and fly with a 15-degree wind correction angle.

Airplane drifts upwind of course Parallel the course, and let the wind push you back on course as depicted in the illustration on page 7 - 14. When back on course, turn 5 degrees into the wind, and fly with a 5-degree wind correction angle. If this is the proper correction, the CDI will remain centered.

Never chase a CDI with large, random heading changes.

Once the wind correction angle is within 5 degrees of that required to track the course, you can *chase* the CDI using small heading changes that never exceed 5 degrees.

Several factors will cause additional drifting: the present heading is not perfect, the wind is changing, the course signal is scalloping, or station passage is imminent. Correct these deviations with small heading changes. At station passage, the VOR receiver's To-From flag changes to From and the CDI will oscillate back and forth like a windshield wiper.

Strong Winds

The first step for tracking a course used a 20-degree intercept angle. In extremely strong wind, that angle must be doubled. You must start with a 40-degree initial correction. At course intercept, remove half of the correction and fly with a 20-degree wind correction angle. The airplane will either remain on course or it will drift downwind of the course.

If you drift downwind, re-intercept the course using the 40-degree intercept, and then fly with a 30-degree wind correction angle. If drifting still occurs, either upwind or downwind of the course, re-intercept the course and use the appropriate wind correction angle, either 35 degrees or 25 degrees. Now that you are within 5 degrees of the proper heading, you can chase the CDI with small heading changes.

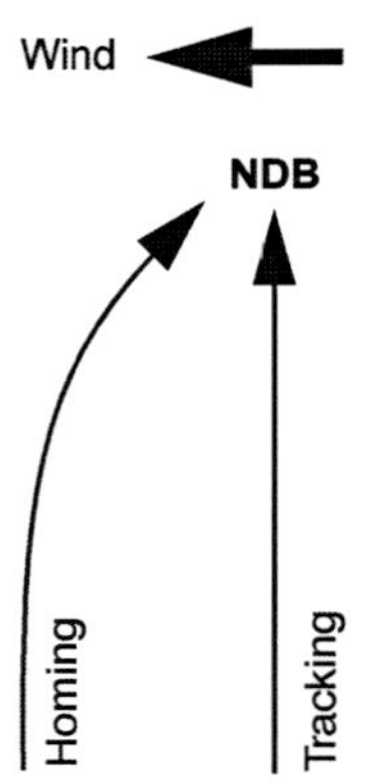

Homing Versus Tracking

ADF HOMING

ADF homing is required for private pilot certification if the airplane you use for the practical flight test has an ADF receiver. To home to an NDB station, turn as necessary to keep the bearing pointer at the top of the radio compass card. Eventually you will reach the station, but if a crosswind exists as depicted by the adjacent illustration, your ground track will be curved, not straight.

ADF TRACKING (Not required for private pilot certification)

To track an NDB course, turn to the no-wind heading when you intercept the course. The deviation angle will now be zero. If drift occurs, turn 20 degrees into the wind. When back on course with a 20-degree deviation angle, remove half of the 20-degree correction, and fly with a 10-degree wind correction angle and a 10-degree deviation angle. One of three things will then happen: the airplane will either remain on course or it will drift downwind or upwind of the course.

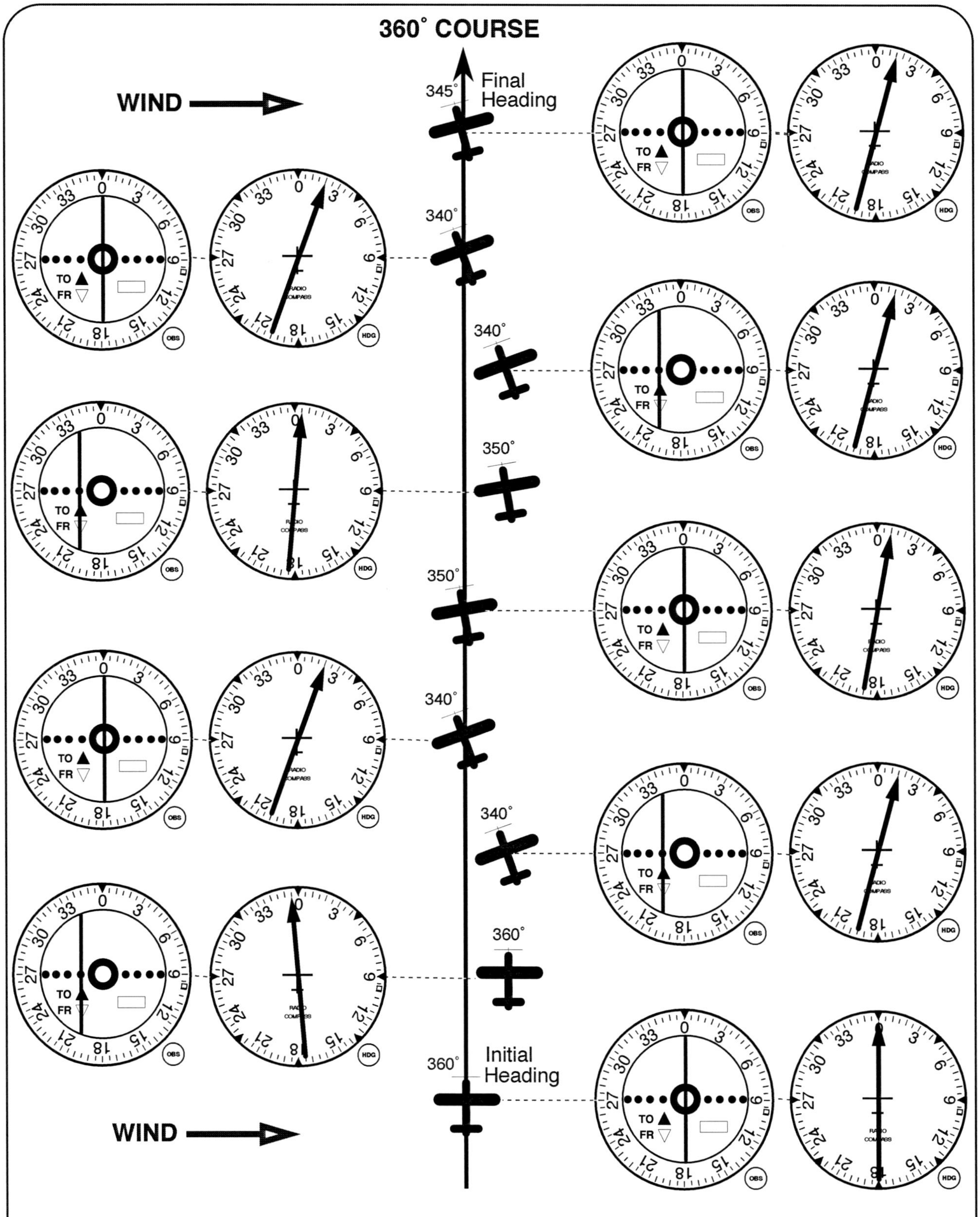

Airplane Drifts Downwind of Course After First Two Heading Changes

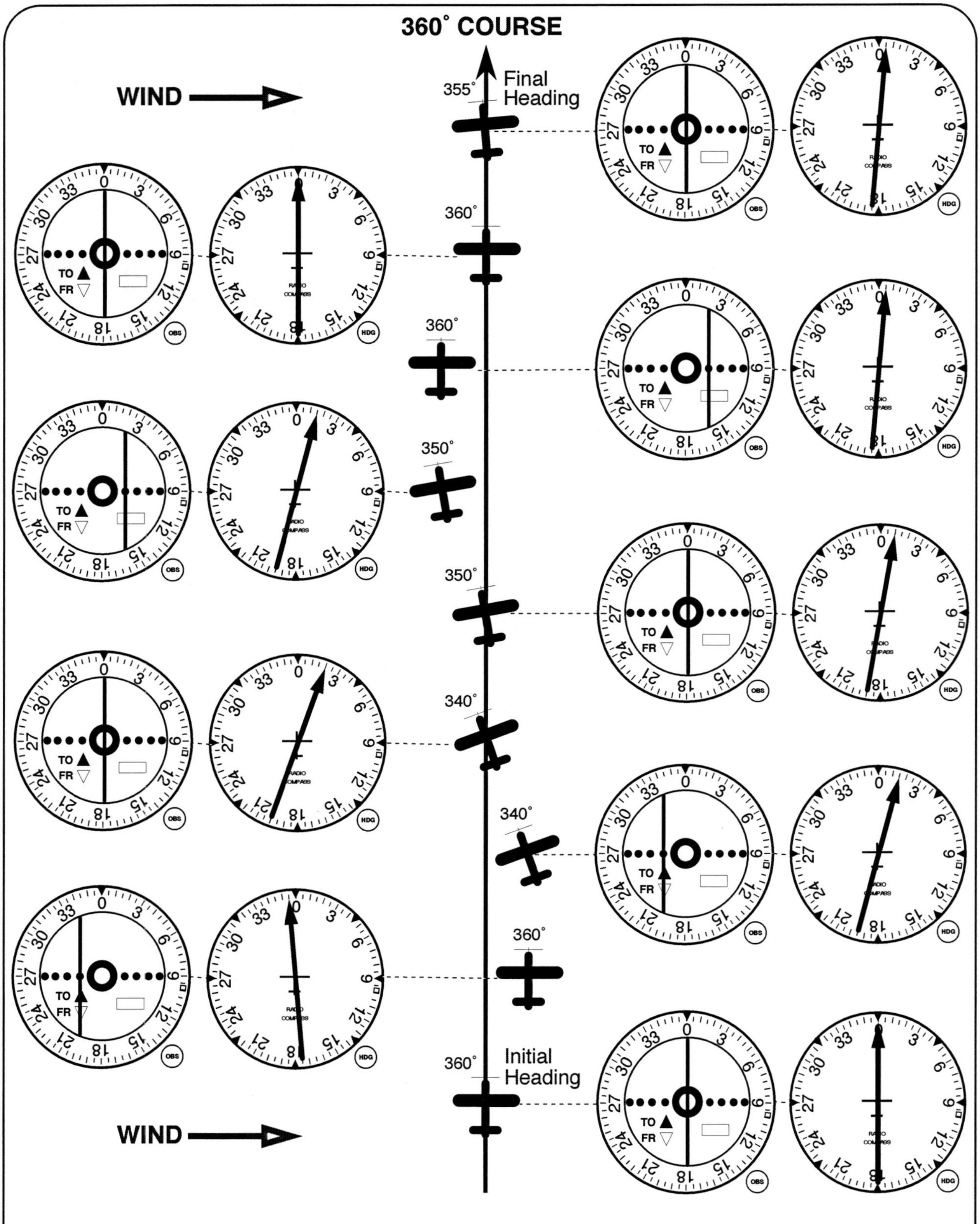

Airplane Drifts Upwind of Course After First Two Heading Changes

Airplane remains on course The deviation angle remains at 10 degrees.

Airplane drifts downwind of course Re-establish the original 20-degree correction as depicted in the illustration on page 7 - 13. When back on course turn 5 degrees toward the course, and fly with a 15-degree wind correction angle and a 15-degree deviation angle. If this is the proper correction, the deviation angle will remain at 15 degrees.

Airplane drifts upwind of course Parallel the course, and let the wind push you back on course (zero deviation angle) as depicted in the illustration on page 7 - 14. When this occurs, turn 5 degrees into the wind, and fly with a 5-degree wind correction angle and a 5-degree deviation angle. If this is the proper correction, the deviation angle will remain at 5 degrees.

REVIEW TEST

1. You tuned in a VOR station that is 80 miles away, and you centered the CDI with a To indication. If you make 360-degree steep turns, how will the CDI and the To-From flag react?
2. To avoid reverse sensing when using VOR equipment, what saying must you remember regarding the airplane's heading and the omni bearing selector?
3. When you select a VOR station, what three things must you accomplish?
4. You are tracking toward a VOR station and notice an off-flag on the OBS, but the CDI is still centered. Can you continue tracking toward the VOR?
5. You tuned in an NDB station that is 80 miles away. If you make 360-degree steep turns, how will the bearing pointer react?
6. When you set the ADF's radio compass card so that zero is at the top, it is said to be in what position?
7. When do you rotate the ADF's radio compass card so that the airplane's heading appears at the top of the card?
8. How can you test ADF equipment to make certain that it is functional?
9. What is the difference between the ADF and the REC or ANT selector positions on an ADF receiver?
10. When you select an NDB station, what three things must you accomplish?

11. You notice an OFF flag on the radio compass card. What does this indicate?
12. You are asked to determine where you are in relation to a VOR station. What is this called and how do you do it?
13. You are asked to determine where you are in relation to an NDB station. What is this called and how do you do it?
14. With ADF equipment, can you determine your position in relation to a commercial broadcast station?
15. How do you know when you are on course with VOR equipment?
16. How do you orient yourself to a VOR cross-bearing?
17. What is the difference between ADF homing and ADF tracking?
18. After the initial VOR or ADF course intercept, you drift off course. What is your initial correction?
19. After you reintercept the course, what action do you take?
20. If the heading you selected in question 19 caused you to again drift downwind of the course, how would you correct your track?
21. If the heading you selected in question 19 caused you to drift upwind of the course, how would you correct your track?
22. When you are within 5 degrees of the proper wind correction angle, how do you correct for subsequent drift indications?
23. What OBS indications occur when you pass over a VOR station?
24. What radio compass indications occur when you pass over an NDB station?
25. How do you know if you have lost a VOR or an NDB signal?

THE BETTER YOU GET,

THE MORE CONFIDENT YOU BECOME.

BEWARE! WHEN CONFIDENCE BECOMES

COMPLACENCY, YOU'RE HEADED FOR TROUBLE.

CHAPTER 8

GPS NAVIGATION, INSTRUMENT AND NIGHT FLYING, AND INSIGHTS

The global positioning system (GPS) validates an old axiom: "Technology simplifies that which was complicated and complicates that which was simple." Even so, with proper knowledge and proficiency, you will enjoy this terrific system and my procedure for learning how to use it.

Private pilot training concentrates on visual flying techniques. Elementary instrument training is required, however, in order to provide protection in the event that you inadvertently fly into clouds or reduced visibility or fly at night over areas with minimal surface lighting.

Night flying can be very enjoyable—smooth air and beautiful visual scenes. With respect to safety, however, night flying reduces your options; and it requires additional knowledge, planning, and preparation.

I have written the monthly *Insights* column in the AOPA's *Flight Training* magazine since 1991. This chapter contains the majority of the articles that I selected for this manual.

GPS

General

The GPS navigation system is now accurate to within 40 feet.

The Global Positioning System (GPS) is composed of three segments: space, user, and control. The space segment consists of 24 NAVSTAR satellites; the user segment consists of all receivers. The NAVSTAR satellites (NAVigation System by Timing And Ranging) orbit Earth every 12 hours at an altitude of 11,000 miles. The system is accurate to within 40 feet.

The control segment consists of a network of ground-based monitoring and control stations: Colorado Springs (the master station), Kwajalein (west of Hawaii), Diego Garcia (Indian Ocean), and Ascension (south Atlantic Ocean).

Each satellite continually transmits almanac data (its approximate position), ephemeris [if·em·er·is] data (its precise position), and the transmission time. The GPS receiver uses this signal to determine its distance from the satellites.

To use GPS navigation, you must receive at least four satellites.

When you turn a receiver on, it performs an internal self test and an initialization test. It must receive signals from at least four satellites (usually six or more are in view) and synchronize its internal clock with the satellite clocks that have nanosecond accuracy (one billionth of a second).

Once the receiver knows its distance from the applicable satellites, it can triangulate its location. Three satellites could do this, but four are required in order to eliminate pseudo-range errors (signal degradation due to the ionosphere, the receiver's clock and processing errors, and poor satellite geometry). VOR triangulation required only two stations, because VOR uses azimuth, not distance.

GPS is not perfect. Navigation integrity—the ability to detect ambiguity in a navigation solution—is the problem. If a local or wide area augmentation system (LAAS or WAAS) is not active, every IFR-certified GPS receiver must continuously perform receiver autonomous integrity monitoring (RAIM), a function that ensures an accurate position. To do this, five satellites must be in view or four satellites and a pressure altitude input from the aircraft's altimeter.

Caution!
Some GPS receivers have functions that seemed worthwhile to the designers, many of whom are not pilots. When pilot workload is considered, these functions are often impractical and do nothing more than complicate the learning and operational process.

Advanced technology equipment does increase a pilot's management capabilities, but there are limits. The cockpit management priorities—in their order of importance—are aviate, navigate, and communicate. For single-pilot operations, these priorities are usually compromised when the workload of GPS management is added.

To keep workload within reason when using GPS navigation, you must let the autopilot fly the airplane.

An autopilot is the solution for this problem. If your airplane has one, you should learn to use it so that it does the aviating while you manage the GPS receiver and maintain an acceptable workload. Otherwise, you will compromise flight safety and your flying skills.

GPS receivers have numerous multifunction controls that are subject to fat-finger errors—activating the wrong input, or activating the correct input at the wrong time. Also, the display screen has many status fields, symbols, and colors that must be carefully monitored. An accidental input error or unobserved status field will result in an operational error.

Pilot rules: Never program GPS equipment while taxiing. Never program in the air unless you can maintain a strict traffic-watch discipline—only a fool would rely strictly on ATC for traffic avoidance. NASA proved that competent pilots flying in visual conditions spend sixteen seconds looking outside the cockpit for every four seconds spent inside the cockpit. Programming a GPS waypoint takes about twenty seconds.

To rely solely on GPS navigation, the database must be updated every 28 days.

GPS equipment cannot be used until the internal self-tests and signal acquisition have been completed and you have verified both your GPS position and the initialization indications as required by the receiver's reference manual. The GPS database revision cycle is every 28 days. If the receiver does not have a current database, you must always confirm your GPS position with radio aids and navigation charts.

A Practical Transition to GPS

GPS navigation is relatively easy to master if you fly frequently and always use the same GPS receiver. If you fly infrequently, however, in different airplanes with different GPS receivers, mastering GPS navigation can be frustrating.

My approach to this problem will simplify this transition. I'll take you from a known system (VOR) to an unknown system (GPS) by using familiar procedures and thought processes. When you master this easy-to-remember procedure, you have established a reliable starting point for using most of the current, aircraft-installed GPS receivers. The manufacturer's reference manual will explain other useful GPS functions.

GPS receivers have two operating modes: Leg and OBS. The OBS mode works just like a VOR receiver.

GPS receivers have two operating modes: Leg and OBS. Both modes depend on specific waypoints—an airport, a VOR, an NDB, a latitude/longitude coordinate, or a bearing and distance from any of those references. In other words, you can quickly make any point on Earth a waypoint for GPS navigation.

The OBS mode always retains the waypoint that you enter or the waypoint that existed when you selected this mode. For a waypoint to change when you pass it, you must have multiple waypoints entered as a flight plan, and you must be in the Leg mode, which is not the subject of this discussion. GPS receivers normally start in the Leg mode, but you need to be in the OBS mode. To do this, you push the receiver's OBS button.

Now, push the receiver's Direct-To button. A highlighted entry field will appear on the display screen, so enter your desired waypoint in that field by using the receiver's control knobs (rotate to enter data) or key pad (type to enter data) as described in the receiver's reference manual.

If an airport has a numeric identifier such as L35, do not use the K prefix if you make it a GPS waypoint.

Lets enter your home airport using its international identifier. If the airport has a three letter identifier, add a K to the prefix—SNA would be KSNA. If the airport has a numeric identifier such as L35, do not use the K prefix. Check the receiver's display screen after you push the Enter button to make certain that the OBS mode is still active.

The switch that allows an HSI or OBS to receive input from either the GPS or a VOR receiver is called the suicide switch.

A specific course will now appear near the OBS annunciation. To change that course, use the receiver's control knobs or an external course indicator such as an HSI or an OBS. Some GPS receivers restrict you to an external course indicator for such changes, and that indicator must be in the GPS mode, not the Nav (VOR) mode. The switch that allows an HSI or OBS to receive input from either the GPS or a VOR receiver is called the suicide switch, because if it is set incorrectly, the results could be deadly.

Now, select the receiver's navigation page that displays a course deviation bar (CDI) and a To-From indicator. The GPS receiver will now work just like a VOR receiver.

Assume that your airport has a Runway-18, which you will depart on and then intercept and track a 225-degree course. Select a 225-degree GPS course and depart. Immediately after takeoff, the CDI will show that the course is to your right and the To-From indicator will show that you are flying away from your airport—the active waypoint—just like a VOR receiver would indicate if a VOR was located at your airport. Turn right and intercept the 225-degree course.

Now, push the Direct-To button and enter another airport's identifier. That airport is now the active waypoint. Check to see that you are still in the OBS mode. Change the course selector until the CDI on the Nav page is centered with a To indication. If the new course is 250 degrees, turn and track that course. Note that the GPS receiver also indicates your groundspeed and distance to the new (active) waypoint.

If the new airport is using right traffic for Runway-04, the traffic pattern's 45-degree entry leg would be course of 265 degrees. When you are within 10 miles of the airport, set the GPS course to 265 degrees and intercept it. You are now on the traffic pattern's entry leg, and the mileage from the airport is continually shown on the display screen.

Caution, a GPS receiver will suck the eyes out of your head. Don't let that happen. You must always scan for traffic.

Just remember that all GPS receivers are equivalent to a powerful vacuum cleaner that will suck the eyes out of your head and glue them onto the receiver's display screen. You're an accident waiting to happen if you let this occur. You must scan for traffic while you are managing a GPS receiver.

GPS Navigation

For radio navigation, you mastered orientation and tracking procedures. GPS navigation eliminates those procedures.

To track a course using GPS, keep the actual track aligned with the desired track.

To orient yourself with GPS, you look at the map on the display screen and immediately see your relative position to any VOR, NDB, or airport that is displayed. To track with GPS, you look at the display screen and turn the airplane so that the actual track equals the desired track. When that occurs, you will stay on course. The wind correction angle is the difference between the actual track and the airplane's heading.

Now, refer to the receiver's reference manual and learn how to enter a flight plan—a series of waypoints—and use the Leg mode, which automatically sequences the waypoints and the desired tracks between those waypoints. It's easy to do once you have mastered the OBS mode in the manner that I just described.

INSTRUMENT FLYING

General

Student pilots must obtain three hours of instrument training in order to qualify for the private pilot certificate. When properly taught, this extremely limited exposure to instrument flight may save your life if you were to unexpectedly lose outside, visual references during a flight.

12 to 15 hours are required to properly teach the first stage of instrument instruction.

To understand just how limited this training is, consider this: A pilot who starts training for an instrument rating will spend 12 to 15 hours completing the first stage of instruction, which consists of instrument scanning, instrument interpretation, and aircraft control. The instrument training described in this section is for emergency use only.

Unfortunately, this limited training develops a bad habit. It teaches you to concentrate on the attitude indicator (AI) and the instruments that give you specific numbers like airspeed, heading, and altitude. When you enroll in a good instrument training program, you will quickly learn that the instruments with the numbers—they're called the primary instruments—are the third scanning priority, *never* the first priority.

Training for emergencies mandates special considerations. First and foremost, the procedures must be simple to execute and easy to remember. Second, they must work in the worst situation that you could encounter and survive.

The lack of surface lighting when flying over sparsely populated areas at night is survivable, providing your altitude is well above all terrain and man-made obstructions. An inadvertent encounter with most clouds is survivable, providing the outside air temperature is at least three to five degrees above freezing. If you can see the glow of your wingtip navigation lights at night, you have entered a cloud.

You will not survive an inadvertent excursion into a cloud that masks terrain or obstructions.

You will not survive an inadvertent excursion into a cloud that masks terrain or obstructions. Survival is also doubtful if you encounter cumulonimbus clouds, clouds that create airframe ice, or freezing rain that falls below the clouds and covers your windshield and airfoils with ice. These grave mistakes represent extremely poor flight planning, poor judgment, and poor utilization of your inflight resources—hazardous inflight weather advisory service (HIWAS), enroute flight advisory service (Flight Watch), and flight service stations (FSS).

Emergency instrument training will make more sense if you understand the five actions that you must master in order to deal with the loss of visual references: cockpit preparation; the 180-degree level, climbing, and descending turns; and recovery from unusual attitudes.

Cockpit Preparation

When you realize that you have or are about to lose visual references, your first task is cockpit preparation. Reduce power to about 2,000 to 2,200 rpm (slow cruise power), trim the elevator in order to fly straight and level for a moment while you set the AI to a level-pitch attitude, align your heading indicator (HI) with the magnetic compass, apply full carburetor heat, and turn the pitot heat on.

If the engine runs rough when you must leave carburetor heat on, lean the mixture until the engine runs smoothly.

Some situations will obviously not require the speed reduction, carburetor heat, or pitot heat. Use good judgment in that respect. If carburetor ice is not present, but the engine runs rough because carb heat must be left on to prevent carburetor ice, lean the mixture until the engine runs smoothly.

After you establish straight-and-level flight and set the attitude and heading indicators, you can start maneuvering. If you reduced power for slow cruise—the power setting you use in the practice area or when on the traffic patterns downwind leg—airspeed will decrease toward maneuvering speed, which protects you in severe turbulence. Carburetor heat prevents carburetor ice and engine failure; pitot heat prevents pitot ice and the loss of your airspeed indicator.

When flying instruments in basic-training airplanes, we refer to the AI's pitch indications in terms of bar-widths as shown in the following illustration. It is important that you set a level-pitch attitude on the AI during the cockpit preparation procedure.

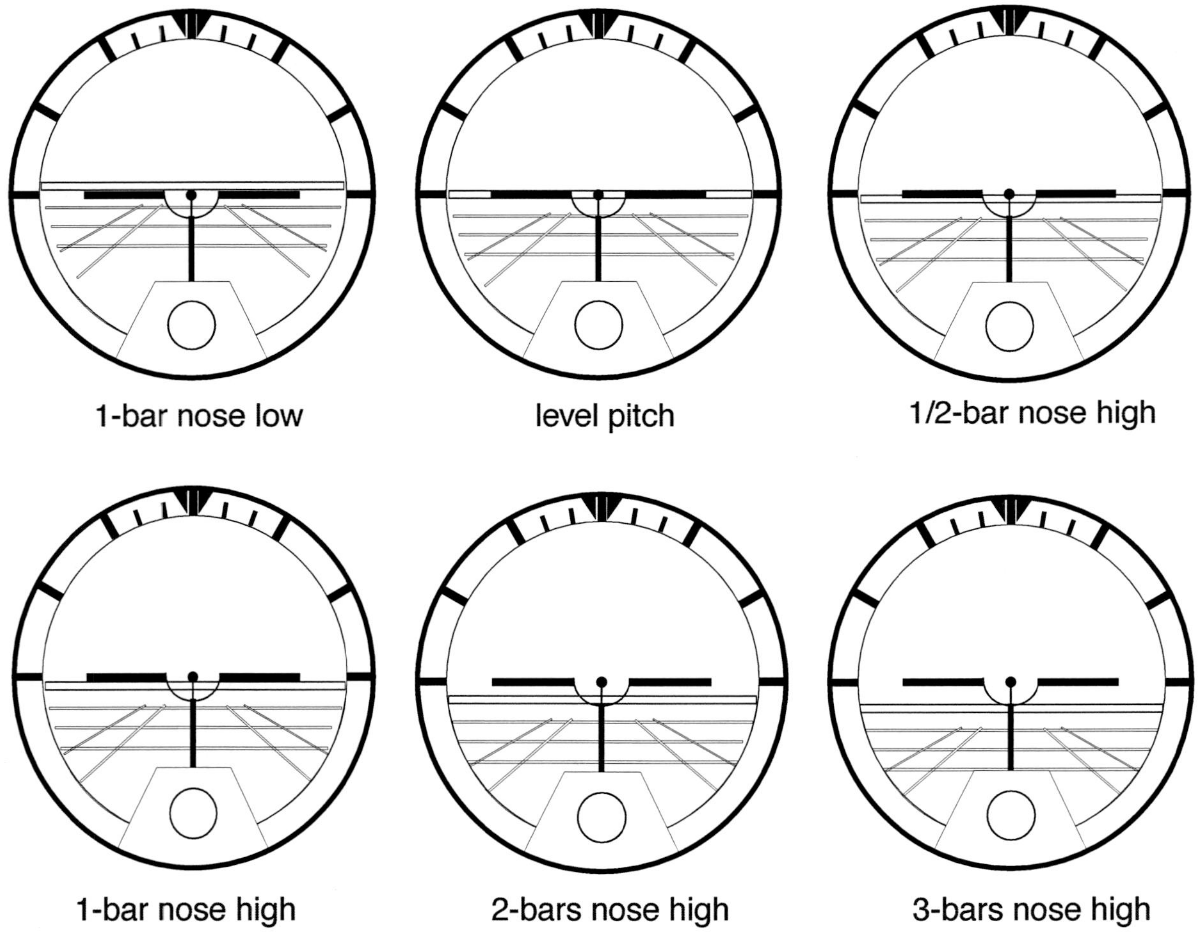

Referencing Pitch Attitudes

Level Turn

Now, if a level, 180-degree turn will restore VFR conditions, stare at the AI, roll into a 15-degree bank, and maintain the level-pitch attitude. Never bank more than 15 degrees when turning. Steeper bank angles make the HI move too fast for proper scanning and make altitude and airspeed control more difficult.

Never bank more than 15 degrees when turning.

Once the bank is established, scan the AI, the altimeter, and the HI. When you approach the desired heading, stare at the AI and roll out of the turn. When the wings are level, scan the AI, altimeter, and HI.

Climbing Turn

If a climbing, 180-degree turn will restore visual conditions, stare at the AI; add full power; establish a wings-level, two-bar nose-high attitude; and turn off the carburetor heat. Check airspeed and return to the AI. If necessary, adjust pitch for the desired airspeed. Never stare at the airspeed indicator while you change pitch attitude. Whenever you change attitude, you must stare at the AI until the new attitude is established.

Never stare at the airspeed indicator while you change pitch attitude.

Now, roll into a 15-degree bank and maintain the climb's pitch attitude. As you approach the desired heading, roll out of the turn. When the wings are level, scan the AI, airspeed indicator, and HI. Continue climbing to visual conditions. If you are lost or unable to reach visual conditions, follow the four-C's rule described on page 6 - 16.

Descending Turn

If a descending, 180-degree turn will restore visual conditions, stare at the AI; reduce power three- or four-hundred rpm; and establish a wings-level, one-bar nose-low attitude. Check airspeed and vertical speed, and return to the AI. If necessary, adjust pitch for the desired airspeed; adjust power for a descent rate of approximately 500 feet per minute.

Now, roll into a 15-degree bank while maintaining the descent's pitch attitude. When you approach the desired heading, roll out of the turn. When the wings are level, scan the AI, airspeed indicator, and HI. Continue descending to visual conditions.

Unusual Attitudes

Recovery from unusual attitudes is the last objective. Put your head down and close your eyes while your instructor places the airplane into an abnormal climbing or descending, steep-bank attitude.

When you're told to recover, look at the airspeed for a split second. If it's increasing, set idle power; if it's decreasing, set full power.

If you try to pull the nose up before you roll out of a diving turn, you may overstress the airplane.

Now stare at the AI. If a nose-high attitude exists, decrease pitch to the level-pitch attitude and roll out of the bank. If a nose-low attitude exists, *roll out of the bank first*, and then increase pitch to the level-pitch attitude. If you try to pull the nose up before you roll out of a high-speed, descending turn, recovery may be impossible and you may overstress the airplane. After you recover, realign the HI with the magnetic compass.

Aircraft Control

Heavy control pressures yield rapid attitude changes that make scanning impossible, because the pressure instruments move too fast. When this happens, concentrate on the AI until those instruments have stabilized.

Light control pressures and slow attitude changes are a prerequisite for instrument flying. To eliminate a tight grip on the yoke and to make sure that the airplane is in trim, momentarily let go of the yoke whenever possible.

Control Corrections

Two-step instrument corrections and two-step control inputs are mandatory for instrument flight.

When flying instruments, instantaneous, one-step corrections are a fallacy, an impossibility, a figment of one's imagination. Attempting one-step corrections constitute a major instrument flying error. All corrections take time, and all require two steps. The two-step corrections discussed on page 3 - 9 are mandatory for both visual and instrument flight, and they are now repeated for your convenience.

Two-Step Control Input Rule First, put in a correction that is obviously excessive, something that you know will work. Second, when it becomes apparent that the correction is starting to work, remove half of the initial input.

Two-Step Correction Rule First, stop the incorrect trend that caused the deviation. Second, establish a new trend that will eliminate the deviation.

Caution!
The instrument procedures that I have just described are for emergency use only. If you want to enhance these skills after you receive your private pilot certificate, purchase my *Instrument Pilot Flight Training Manual* and enroll in a good instrument training program. Upon completion, you'll realize the inadequacy of three hours of instrument flight training and the fallacy of placing total dependence on the AI and the primary instruments—a grossly misleading term that causes a complete misconception of instrument flying.

Primary instruments is a grossly misleading term that causes a complete misconception of instrument flying.

NIGHT FLYING

Never fly at night with a weak airplane battery.

- Never fly at night if the airplane battery—your emergency electrical supply—is weak. Engine start gives a good indication of battery condition. Does the starter readily crank the engine, or does the starter labor? After engine start, is the ammeter charge rate normal or abnormally high? A prolonged, high-charge rate indicates a weak battery.

- Plan a route that gives you the most forced landing options. For those airports, check the hours of operation and whether or not they have pilot controlled lighting.

- Always carry two flashlights and spare batteries. Flashlight technology and miniaturization is constantly improving, but as a minimum, your flashlights should have the candlepower of the old flashlights that used two D-cell batteries.

- Never place a flashlight near the magnetic compass.

- Check the operation of all aircraft lights prior to flight.

- Current federal regulations create confusion regarding the use of exterior anticollision lights because of the various lighting configurations that exist on modern airplanes. In September, 2003, the FAA published Advisory Circular 91-73A, which clarified this situation somewhat.

Never use the strobe lights during taxi if they will adversely affect the vision of other pilots or ground personnel.

- Aircraft lights may be used to make an aircraft operating on the airport surface at night more conspicuous. Pilots may now use various combinations of exterior lights to convey their location and intent to other pilots. Strobe lights should never be on during taxi if they will adversely affect the vision of other pilots or ground personnel.

- Engine running: Turn on the rotating beacon (if installed).

- Turn on all anticollision and landing lights when taxiing onto the runway if cleared for takeoff. If cleared into position and hold, do not turn the landing lights on until you receive your takeoff clearance.

You cannot always trust your vision when flying. Many landing accidents have been attributed to the deceptive nature of human vision in certain conditions. Here are some examples:

- A wider than normal runway appears closer than it is.

- A narrower than normal runway appears farther away than it is.

- An up-sloping runway creates the illusion of being high.

- A down-sloping runway creates the illusion of being low.

- Haze, smoke, dust, darkness, and precipitation create the illusion of being high.

- Rain on the windshield or windshield defects that refract light can create the illusion of being high. This is particularly evident on dark, rainy nights.

- Viewing an airport through an intervening rain shower makes the runway lights look bigger and creates the illusion of being high.

- Extremely clear air makes objects seem closer than the actual distance.

- Bright lights appear closer than the actual distance, and dim lights appear farther away than the actual distance. This effect is greatly enhanced if surface lighting is minimal.

INSIGHTS (More of my articles from the AOPA's *Flight Training* magazine)

The following articles should make you realize these facts:

- Do not make flying complicated. Keep it simple.
- Attitude, proficiency, and knowledge are your primary concerns. That's what keeps you safe.
- Flying is an addictive endeavor.

Thingamajigs And Other Stuff

Do you know your airplane? - May 2001 issue

When evaluating a flight instructor for employment, I open an engine cowling and ask the applicant to identify the major engine components. Many of them are unable to do this, and that includes some who have graduated from college-level aviation training programs.

For example, they can't identify the exhaust manifold—that rusty looking array of plumbing that is an important preflight inspection item—and they have no idea why those dirty rubber baffles sit atop the cylinders—a preflight item that is critical for proper engine cooling.

I feel sorry for these pilots because as a teenager, I had a deep interest in mechanics, and it was easy to obtain an old car, tear the motor apart, rebuild it and make it run. That opportunity has become almost non-existent.

Many pilots acquire rote mechanical knowledge, but they have trouble applying it to practical situations. When a cockpit problem occurs, they may make an assumption that results in an incorrect course of action.

Years ago the airlines stopped teaching the nuts and bolts of aircraft systems, and I am glad they did. The complexity of some systems had become ludicrous and had no bearing on a pilot's ability to analyze a problem and take corrective action. Today, an airline pilot must know how each system functions, how to deal with component failures, how to interpret each warning light and instrument indication, and how to confirm that the desired action did, in fact, occur when a switch was moved or a control was activated.

To acquire the proper mechanical knowledge for flying, start with the emergency section of the pilot's operating handbook (POH). Learn all you can about the systems mentioned.

The electrical system is a good example. What is wrong if the alternator warning light comes on during flight? This light is nothing more than a voltmeter. When it's off, system voltage equals alternator voltage and all is well. When it's on, system voltage equals battery voltage, and the alternator is not supplying power.

The POH will tell you to reduce electrical loads and turn the alternator switch off and then on, but before you do this, be sure to turn the avionics equipment off because a damaging voltage spike (momentary high voltage) can occur when the alternator is turned on again. The potential for a voltage spike exists because after you turn the alternator on it takes a microsecond before the voltage regulator starts regulating.

If the alternator has failed, must you land immediately? That depends on your situation, but for visual, daytime flying the answer is not necessarily. Battery life is the primary concern. A fully charged 30-amp-hour battery will provide 30 amps for one hour. I would be pessimistic (an important pilot trait), and I would assume that my 30-amp-hour battery is now a 15-amp-hour battery. If the electrical system is drawing seven amps, the battery should last about 30 minutes.

On the other hand, imagine that your flying over a desolate area. In that case, I would turn the battery switch off in order to save the remaining battery power and have it available when I approach a suitable destination. The engine keeps running because it's powered by magnetos, but all electrical items in the airplane are now dead— lights, avionics, turn coordinator, fuel gauges, oil temperature gauge, etc.

Here is an example of where a lack of system knowledge allows the pilot to make an assumption that leads to an incorrect course of action: The alternator has failed, the pilot has reduced electrical loads, but he has flown for almost 30 minutes. Battery life now concerns him, so he turns off the battery switch to save battery power. When he finally approaches a suitable destination, he turns the battery switch back on, but nothing happens. The entire system is dead. With proper knowledge, he would have known that a relay is activated when the battery switch is turned on, but only if enough battery voltage is available. Otherwise, the battery relay will not activate, and the remaining battery power will be unavailable.

As pilots gain experience, they begin to realize the importance of the mechanic who bends the wrenches, and I've never met a mechanic who wasn't interested in furthering a pilot's knowledge of the aircraft. Therefore, if your mechanical knowledge is not up to par, I would strongly suggest that you hire a mechanic and have him or her explain the importance of everything that sits inside your airplane's engine cowling and on your instrument panel. It's an investment that pays enormous dividends when you are faced with an inflight problem.

Pretzel Pilots

Ground track, what's that? - April 2001 issue

A student pilot enters a traffic pattern's downwind leg too close to the runway. By the time he is ready to turn base, his distance from the runway's extended centerline has increased to more than a mile. The subsequent base and final legs are an amazing series of twists and turns. This guy's a pretzel pilot because he cannot plan and maintain the proper ground track.

I query the student's instructor even though I know what I'll probably hear: "He never does that with me.... I teach ground reference maneuvers in the pattern.... I solo my students in minimum time.... He knows how to fly a constant heading."

When an instructor tells me that his student never does what I see, I want to hand the instructor a gold-plated barnyard shovel. The instructor has probably never sat back and allowed his student to make mistakes. He probably talks nonstop, cautioning the student about what will happen if certain actions are not taken.

That's the bane of modern flight training where students and instructors use headsets, boom mikes, and intercom systems. The instructor who talks all the time is an obstacle to learning. To teach ground reference skills in the traffic pattern is an absurdity. There's too many other things that must be accomplished and too many distractions. Proper ground reference skills cannot be mastered in a traffic pattern, particularly at a busy airport.

Prior to concentrated traffic pattern work and the first solo flight, all students should master tracking over a road, tracking parallel to a road, rectangular patterns, and S-turns across a road. These maneuvers cannot be taught effectively unless wind is present. Consequently, instructors must watch for proper conditions and schedule their students accordingly.

No instructor wants to prolong a student's time to solo, but training priorities should never be compromised in order to minimize that time. I sincerely believe that instructors who do otherwise are just trying to make themselves look good in the eyes of their peers.

Minimum solo time like minimum certification time means nothing to me. I want to see a properly trained pilot who has truly mastered each required skill.

The ability to fly a constant heading does not mean that the student can fly a specific ground track, because heading control does not require wind and ground-track awareness. Show me a pilot who is not constantly aware of his position and the general wind direction and velocity, and I'll show you an improperly trained pilot. I can say the same for the pilot who turns to a planned heading and assumes he'll stay on course.

It doesn't matter if it's a traffic-pattern leg or a cross-country leg. Specific procedures must be followed. When entering a traffic pattern, pick a line on the ground that you want to track. Obviously, specific lines are seldom present, but specific references are—structures, intersections, or other prominent features. Consider wind and turn to a heading that will allow you to track that line. The student who thinks in this manner can easily fly rectangular traffic patterns.

On cross-country flights, momentarily turn to the no-wind heading when you start a new leg. Look straight ahead, pick a distinctive reference point that is on or near the horizon, and pick one or two references points that lie between that point and your present position. Now turn to your planned heading based on forecast winds and monitor the reference points you selected. Adjust your heading as necessary in order to track the desired course and locate the landmarks you circled on your navigation chart during flight planning.

In both cases—the traffic pattern and the cross-country leg—you now spend the majority of your time looking outside the cockpit at ground references in order to evaluate wind drift and ground track.

Do me one big favor and add air traffic to your scan. Thank you. You are now leaving the realm of a primordial airplane driver and advancing toward the exalted position of airplane pilot.

Idiot's Anonymous

Aviation's bad boys! - February 20002 issue

Place me in or near an airplane and like most good pilots I become an extreme pessimist. In other circumstances, I am an eternal optimist, well aware of the voluminous safety and procedural information published monthly by commercial and government sources. I would like to think that all pilots heed this information and use common sense, but unfortunately this is not the case.

I'll never forget the pilot who purchased a beautiful light twin from an aircraft dealership where I worked. An instrument rating from our school was part of the deal, but after the pilot had received about 15 hours of instrument training from one of my best instructors, he stopped scheduling lessons. About three months later, the FAA came in and requested the pilot's training records. He had loaded his family aboard the airplane, filed an IFR flight plan, and attempted an IFR flight to an airport in mountainous terrain near a ski resort. The airplane wreckage was found during the spring thaw.

Another pilot that I knew was cantankerous and extremely opinionated. He owned a single-engine, high-performance airplane, and he acted like God's gift to aviation. On his last cross-country flight, he ran out of fuel about 50 miles from his destination and did not survive the forced landing.

Brilliant pilot number three learned to fly and became obsessed with aviation. With minimum time in his log book and truck loads of cash, he obtained his instrument rating and immediately purchased a high-performance single-engine airplane. That lasted about six months. Against all advice from those who knew him, he sold that airplane and purchased a medium-size twin turboprop. That, too, lasted only six months. He was forced to sell the airplane because no one would fly with him, including his family. Fortunately he's still alive.

My most recent encounter involved a pilot who had flown a single-engine, fixed-gear airplane for several years. After he sold that airplane and purchased a single-engine retractable-gear airplane, I was asked to fly with him "a couple times" for a checkout.

He started our first session by telling me stories about his harrowing experiences in his previous airplane. When my turn came—I don't mind getting paid to hear stories—I started asking him basic questions about the new airplane, none of which he could answer. I asked if he had read the POH, and he said, "I've thumbed through it." Needless to say, we did not fly that day. I spent the time reviewing the POH, cockpit orientation, and the preflight inspection.

An improvement in knowledge was not evident at our next session, but as a favor to the person who asked me to do this, I told the pilot to preflight the airplane. Personally, I had a feeling that this would be my first and last flight with this individual.

I monitored his cockpit setup and preflight inspection, and when complete I asked if he was going to drain the fuel sumps. He said, "I don't have a fuel strainer." When I said, "No drainee, no flyee," he drove to the pilot supply shop and bought a strainer.

We reviewed the typical checkout maneuvers, performed the emergency landing gear extension procedure, and returned for landings. On his first attempt, he abruptly increased the pitch attitude when 20 feet above the runway which resulted in a rapid loss of airspeed and a dangerously high sink rate. He was surprised when I took the controls and executed a go-around, because "everything looked okay" to him. On the next approach, I talked him through a poorly executed landing, terminated the flight, and stated that he would need far more than the five hours of instruction that his insurance company required.

Two days later, while taxiing out for a flight with a student, I heard the same pilot abort a takeoff on the ground control frequency—his airplane had a weird audio panel, the operation of which I had explained and demonstrated. I keyed the mike and asked if he had an instructor with him. After a long pause, he said, "No." I then reminded him and everyone else on the frequency that he was not checked out in the airplane. Sheer insanity!

We all have a responsibility to personally challenge any pilot who does not uphold the standards that are dear to the hearts of most pilots—standards that put aviation safety first and foremost. You should never encourage or fly with such individuals. To do otherwise is just asking for trouble.

Pool-Cue Flying

Aim your plane! - June 2002 issue

Everyone has played pool at one time or another, and I remember my first attempt? I took the pool cue, placed it in line with the track that I wanted the cue ball to follow, and took my shot. Later on, I discovered that if I treated the airplane's wing like a pool cue, I could easily pick out a ground track that I wanted the airplane to follow. Flying ground reference maneuvers and rectangular traffic patterns then became a simple task.

These maneuvers begin with tracking over a road and tracking parallel to a road when a crosswind is present. You learn to compensate for wind drift by making a coordinated turn into the wind and establishing a wind correction angle that will keep you on the desired ground track, either over the road or over a line that's parallel to the road.

When tracking parallel to a road, most instructors will tell you to keep the road in the same position relative to the wing or the wing strut. Yes, that will keep you parallel. However, to be successful with more advanced maneuvers, make your primary reference the line that defines your desired ground track. To do this you must visualize that line on the ground by using specific reference points such as buildings, road intersections, or natural features. If you make certain that the airplane stays on the desired ground track, your ground reference maneuvers will be perfect.

Your next challenge is rectangular patterns. You now combine straight segments with turns, and because you are flying parallel to each side of the rectangle—defined by roads or other linear references—it is easy to pick out the reference lines that define the desired ground track.

S-turns across a road are next. No straight lines here. You are constantly turning, and you learn to control a curved ground track by varying your bank angle as groundspeed changes. This maneuver is easy to fly if you concentrate on crossing the road with the fuselage perpendicular to the road just as the airplane passes through a wing's-level attitude as you roll from a left bank to a right bank or vice versa.

You are now ready for the airport's traffic pattern. That means concentrated takeoff and landing practice, but don't get too excited. Your first priority is to master the traffic pattern. Warning: If your instructor says, "We can skip basic ground reference maneuvers because I can teach you that stuff in the traffic pattern," change instructors.

To start, you takeoff and climb to within 300 feet of pattern altitude on the departure leg and turn onto the crosswind leg (AIM 4-3-4). Before you turn, clear the airspace in the direction of turn and look where your wing is pointing. It is showing you where your crosswind-leg ground track should be on the Earth's surface. If you turn toward those references, your crosswind leg will be perpendicular to the runway.

Use the same procedure for turning onto the downwind, base and final legs. Caution: Before you clear the airspace in the direction of turn, you must *always* clear the airspace on the other side and in front of your airplane. Never assume that every pilot will follow correct procedures or that ATC will advise you of all traffic conflicts.

Last but not least, is wind compensation. You must apply a wind correction angle in order to remain over your desired ground track. If you maneuver the airplane so that it overflies the ground track's reference points, the wind correction angle will be established automatically.

You must then consider the wind correction angle when you use the wing as a pool cue. To pick out a ground track that is parallel or perpendicular to the runway, look to where the wing would point if no wind correction was required. Your desired ground track will lie either slightly ahead of or slightly behind the wing.

Never use the heading indicator when flying ground reference maneuvers. You must constantly look outside the cockpit for attitude and ground track deviations and for traffic. Periodically, you can briefly glance inside the cockpit to check airspeed and altitude.

Wind and ground track are the nemesis of every poorly trained pilot. Make wind awareness and ground-track planning a top priority. And remember, wind awareness starts when you walk out to the airplane.

Say It Isn't So

Wisdom that can't be ignored - February 2003 issue

The countdown has begun. In 12 months, on December 17, 2003, numerous centennial celebrations will be held in honor of Orville and Wilbur's first powered flight. The technological advancements during this period have been nothing short of spectacular, but I pity the pilot who ignores the numerous pearls of wisdom that experienced pilots generated during this era. My favorite "pearls" follow, and if I know the author's name, it is listed.

"Rule one: No matter what else happens, fly the airplane."

"There are old pilots, and there are bold pilots, but there are no old, bold, pilots!"

"Aviation safety consists of attitude, proficiency and knowledge." (R. C. Morton, FAA safety program manager, retired.)

"Professionalism is like pornography, you'll recognize it when you see it." (John Lauber, National Transportation Safety Board, retired.)

"A professional is someone who can do his best work when he doesn't feel like it." (Alistair Cooke)

"Truly superior pilots are those who use their superior judgment to avoid those situations where they might have to use their superior skills."

"Good judgment comes from experience and experience comes from bad judgment."

"Always know where you've been, where you are, and where you're going."

"Always remember, you fly an airplane with your head, not your hands."

"The only time you have too much fuel is when you're on fire."

"An airplane may disappoint a good pilot, but it won't surprise him."

"Don't drop the aircraft in order to fly the microphone."

"An airplane flies because of a principle discovered by Bernoulli, not Marconi."

"Learn from the mistakes of others, You won't live long enough to make all of them yourself."

"The three most useless things in aviation: the altitude above you, the runway behind you, and the fuel that you did not put in the tanks."

"Any pilot who relies on a terminal forecast can be sold the Brooklyn Bridge. If he relies on winds-aloft reports, he can be sold Niagara Falls."

"Weather forecasts are horoscopes with numbers."

"Any attempt to stretch fuel is guaranteed to increase headwind."

"It's easy to make a small fortune in aviation. You start with a large fortune."

"A fool and his money are soon flying more airplane than he can handle."

"Remember, you're always a student in an airplane."

"Keep looking around; there's always something you've missed."

"Never trade luck for skill."

"Airspeed, altitude and brains: At least two are always needed to successfully complete a flight."

"Flashlights are tubular metal containers kept in a flight bag for the purpose of storing old dead batteries."

"Flying the airplane is more important than radioing your plight to a person on the ground incapable of understanding it."

"Stay out of clouds. The silver lining everyone keeps talking about might be another airplane going in the opposite direction."

"You start with a bag full of luck and an empty bag of experience. The trick is to fill the bag of experience before you empty the bag of luck."

"Instrument flying is when your mind gets a grip on the fact that there is vision beyond sight." (Captain Pettibone, the U.S. Navy's fictitious safety officer, in a 1945 issue of *Approach* magazine.)

"The emergencies you train for almost never happen. It's the one you can't train for that kills you." (Ernest K. Gann)

"What is the similarity between air traffic controllers and pilots? If a pilot screws up, the pilot dies; If ATC screws up, the pilot dies."

The Big Lie

Accident avoidance - June 2003 issue

Accidents happen. That's a fact of life. As a flight instructor, I demonstrate one of two poor-judgment situations—buzzing or scud running—to students who are in their final stage of private-pilot training.

Buzzing occurs when you make a high-airspeed, high-rpm dive that terminates at low altitude. The high velocity of the propeller's blade tips generate a noise that causes everyone on the ground to look up.

If we are flying in the early morning or late afternoon, I orient the maneuver so that I am diving toward the rising or setting sun, which makes it impossible for the student to see the large radio tower that lies ahead of us. When just below the top of the tower, I say "Watch out!" and abruptly climb away from the hazard. Students never see the tower until I make the callout.

Scud running occurs when you attempt to fly below low, terrain-obscuring clouds in order to maintain visual conditions. You are trying desperately to get to your destination, and it looks like you can make it.

If we are flying when it's hazy and the lighting is correct, I tell the student to assume that low clouds are present, and we are going to fly at low altitude through a pass in order to get home. The student does not know that a group of hard-to-see radio towers are located on the far side of the pass. When we get close to the tower's, I say "Watch out!" and abruptly climb away from the hazard. Students seldom see the towers until I make the callout.

In each case, I tell students that buzzing and scud running are two of the dumbest things a pilot can do. Those actions cause accidents.

Everyone then says, "Okay, but I would never do that."

And I say, "That's the big lie. The pilots who had the accidents made the same statement when they were student pilots."

Flight safety experts have beaten human factors and accident prevention to death. You are exposed to their findings in every flight training course, aviation seminar, and aviation publication. Even so, accidents continue.

Flight Facts, a U.S. Army aviation publication made this statement: "When first attained, each new plateau of risk seems to be the last." Compare how you felt during your first night flight to how you felt after making several such flights. The same comparison can be made for flying over mountainous terrain. For every repeated endeavor, the risk factor will decrease in your mind, but in reality it never changes. The assumption that risk has diminished is called *pilot complacency,* and this is a major factor in many aviation accidents.

A retired FAA Air Safety Program Manager said, "Flight safety consists of attitude, knowledge, and proficiency." That's the best summary I've ever heard, and pilot complacency would most certainly fall under the attitude category.

I've observed other attitudes that cause problems. When I see a pilot who acts as an aviation know-it-all or acts as though he is God's gift to aviation, I see an accident waiting to happen. Most of the outstanding pilots that I've flown with over the years don't flaunt their talents or knowledge. They know how easy it is to make a mistake or fall under the spell of the big lie. Aviation is no place for big egos.

During December 2002 and January 2003, three fatal accidents occurred in my area. Two were the result of scud running and the third was structural failure while the aircraft was vectored for an instrument approach. Thunderstorms had been forecast and were, in fact, in the area—yes, they can occur during the winter months. These three accidents involved the big lie, complacency, and failure to use available resources.

You must know how to interpret weather reports and charts—now available on the internet; how to obtain a proper weather briefing; and how to utilize flight service stations, enroute flight advisory service (Flight Watch), and hazardous in-flight weather advisory service (HIWAS). Above all, you must never become a victim of the big lie.

Aviation Reality

Enjoy and appreciate what you're flying - November 2003 issue

I don't know who wrote the following story, but all pilots should read it.

"One fine hot summer's afternoon saw a Cessna 150 flying in the pattern at a quiet country airfield. The instructor, bothered with the student's inability to maintain altitude in the thermals, was getting impatient at sometimes having to take over the controls. Just then he saw a twin-engine Cessna 5,000 feet above him and thought, 'Another 1,000 hrs of this and I qualify for the twin charter job! Aaahh, to be a real pilot... going somewhere!'

"The Cessna 402 was already late and the boss told him this charter was for one of the Company's premier clients. He'd already set MCP and the cylinders didn't like it in the Summer day's heat. He was at 6,000 feet with a 20 kt. headwind. Today was the 6th day straight and he was pretty tired of fighting these engines. Maybe if he got 10,000 feet out of them the wind might die off... geez those cylinder temps! He looked out momentarily and saw a B737 leaving a contrail at 33,000 feet in the serene blue sky. 'Oh man,' he thought, 'my interview is next month. I hope I just don't blow it!' Outa G/A, nice jet job, above the weather... no snotty passengers to wait for... aahhh.

"The Boeing 737 bucked and weaved in the heavy CAT at FL330 and ATC advised that lower levels were not available due traffic. The Captain, who was only recently advised that his destination was below RVR minimums, had slowed to LRC to try and hold off a possible in-flight diversion and arrange an ETA that would hopefully ensure the fog had lifted to CAT-II minima. The Company negotiations broke down yesterday and it looked as if everyone was going to take a pay cut. The F/O's will be particularly hard hit as their pay wasn't anything to speak of anyway. Finally deciding on a speed compromise between LRC and turbulence penetration, the Captain looked up and saw a Concorde at Mach 2+. Tapping his F/O's shoulder as the 737 took another bashing, he said 'Now THAT'S what we should be on... huge pay... super fast... not too many routes... not too many legs... above the CAT... yep! What a life...!'

"FL590 was not what he wanted anyway and he considered FL570. Already the TAT was creeping up again and either they would have to descend or slow down. That rear fuel transfer pump was becoming unreliable and the F/E had said moments ago that the radiation meter was not reading numbers that he'd like to see. The Concorde descended to FL570 but the radiation was still quite high even though the Notam indicated hunky dory below FL610. Fuel flow was up and the transfer pump was intermittent. Evening turned into night as they passed over the Atlantic. Looking up, the F/O could see a tiny white dot moving against the backdrop of a myriad of stars. 'Hey Captain,' he called as he pointed, 'must be the Shuttle.' The Captain looked for a moment and agreed. Quietly he thought how a Shuttle mission, while complicated, must be the be all and end all in aviation. Above the gunk, no radiation problems, no fuel transfer problems... aaah.

"Discovery was into its 27th orbit and perigee was 200 feet out from nominated rendezvous altitude with the commsat. The robot arm was virtually unusable and a walk may become necessary. The 200 foot predicted error would necessitate a corrective burn and Discovery needed that fuel if a walk was to be required. Houston continually asked what the Commander wanted to do, but the advice they proffered wasn't much help. The Commander had already been 12 hours on station sorting out the problem and just wanted 10 minutes to himself. Just then a mission specialist, who had tilted the telescope down to the surface for a minute or two, called the Commander to the scope. 'Have a look at this Sir, isn't this the kinda flying you said you wanted to do after you finish up with NASA?' The Commander peered through the telescope and cried, 'Ooooohhhhh yeah!' Now THAT'S flying! Man, that's what its all about! I'd give anything just to be doing THAT down there! What the Discovery Commander was looking at was a Cessna 150 in the pattern at a quiet country airfield on a nice, bright sunny afternoon."

— — — — —

That story illustrates the intrinsic love of flight that motivates and inspires all of us, from student pilot to astronaut. It reminds us that what we do is special, regardless of what we fly. And it certainly enhances my favorite saying: A private pilot's certificate is the best bargain in the USA.

REVIEW TEST

1. What does a GPS receiver do initially when you turn it on?
2. How many satellites must be received for basic GPS operation?
3. What is receiver autonomous integrity monitoring (RAIM)?
4. How many satellites must be received for RAIM operation?
5. What are GPS receiver “fat-fingered” errors?
6. What are the two GPS modes?
7. Which GPS mode works exactly like a VOR receiver?
8. When using the GPS’s OBS mode, what is the reference point?
9. What is the danger in using a GPS receiver for navigation?
10. Why do I refer to private pilot instrument training as emergency instrument training?
11. You encounter instrument conditions. What is the cockpit preparation procedure?
12. What is meant by a two-bar nose-high or a one-bar nose-low attitude?
13. How do you make a level turn using instrument reference?
14. How do you make a climbing turn using instrument reference?
15. How do you make a descending turn using instrument reference?
16. How do you recover from a decreasing-airspeed-climb unusual attitude when using instrument reference?
17. How do you recover from an increasing-airspeed-descent unusual attitude when using instrument reference?
18. What is a prerequisite for instrument flying?
19. What is the two-step control rule?
20. What is the two-step correction rule?
21. Never fly at night if your emergency electrical supply is weak. What does that mean?
22. At night, what is your primary concern for route planning?
23. What is true with respect to flashlights during night flights?
24. When should strobe lights never be operated?
25. At night, will a wide runway appear closer or farther away than it actual is?
26. At night, will an up-sloping runway create the illusion of being high or low?
27. At night, will extremely clear visibility make objects seem closer or farther away?
28. At night, will dim lights appear closer or farther than their actual distance?

CHAPTER 9

COMPLEX, HIGH-PERFORMANCE AIRPLANES

After earning your private pilot certificate and gaining some practical experience, you will get the urge to fly something more exotic. A complex airplane has retractable landing gear, flaps, and a controllable-pitch propeller; a high-performance airplane has more than 200 horsepower.

These airplanes fly faster, carry more cargo and passengers, and require an airplane checkout. However, before you can fly them solo, your flight instructor must give you the applicable logbook endorsement—one for complex airplanes and one for high performance airplanes.

A good checkout requires considerable ground and flight time.

A good checkout requires considerable ground and flight time because several new factors must be mastered: operating limitations, normal and emergency procedures, systems, avionics, performance, and weight and balance. This chapter discusses the elements that must be memorized and those that can be confusing or overlooked.

The necessity for maintaining pilot proficiency increases as airplane complexity increases. If financial or time limitations make this impossible, do not fly these airplanes.

HEADWORK

LIMITATIONS

Airspeed Limitations Memorize the following speeds: best-rate climb, cruise climb, maneuvering, best glide, landing gear extension, flap extension, and final approach with full flaps extended. Refer to the POH for other seldom-used speeds.

Some airspeeds should be decreased as gross weight decreases.

The published best glide, stall, maneuvering, and final approach speeds are valid at the airplane's maximum gross weight. For every two percent below gross weight, these speeds should be reduced approximately one percent. For example, the airplane's gross weight is 4,000 pounds and the maneuvering speed is 130 knots. Two percent of the gross weight is 80 pounds; one percent of the speed is 1.3 knots. For every 80 pound reduction in gross weight, you should reduce speed 1.3 knots.

It is more practical to think in ten-percent weight increments and five-percent speed increments. In the last example, ten percent of the gross weight is 400 pounds; five percent of the speed is 6.5 knots. For every 400 pound reduction in gross weight, you should reduce speed 6 knots.

Compared to fixed-gear airplanes, complex airplanes have less drag and require more time to decelerate from cruise to maneuvering speed. If you encounter or anticipate severe turbulence, reduce power and decelerate to maneuvering speed. If turbulence is only moderate, you can increase pitch attitude somewhat to help decelerate. In extreme cases you can extend the landing gear after you decelerate to the gear extension speed, but if you inadvertently extend the landing gear above the gear extension speed, leave it down for the remainder of the flight. Damaged gear doors or door actuators could cause binding and failure of the normal and the emergency gear extension systems.

Never extend the flaps when flying in severe turbulence.

When flaps are extended, the allowable G-load for many normal category airplane decreases from 3.8 to 2.0 Gs. For this reason, you should never extend the flaps when flying in severe turbulence.

Try to avoid landing gear and flap extension when you are within ten knots of the maximum extension speed. That will reduce wear and tear on the airplane, and with proper planning, it is easy to do.

Do not assume that power must be reduced during a climb.

Powerplant Limitations Memorize the following power settings: takeoff, maximum continuous, and cruise climb. Do not assume that power must be reduced during a climb. Some engines require full throttle for climb so that extra fuel is supplied to the cylinders for valve cooling. Check the POH.

Check for unauthorized combinations of manifold pressure and rpm that lie within the range of normal power settings. These transient-only power settings are unusable because of unacceptable crankshaft and propeller vibration patterns.

Check for minimum oil temperature limits.

Check for minimum oil temperature limits. Some POHs will list a minimum temperature for engine runup or takeoff.

Full seats and full fuel tanks may be illegal.

Weight limitations Memorize the airplane's gross and useful weights. Most six-place, high-performance airplanes have considerable cabin space, but if all seats and fuel tanks are full, the gross weight will be greatly exceeded. This indicates versatility, not poor design—the airplane can accomplish more than one objective. With a large load of passengers or cargo, range must be reduced, or with a reduced load, maximum range can be obtained.

Memorize the airplane's basic empty weight and, if applicable, the maximum zero-fuel weight—the maximum allowable weight of the airplane when loaded with everything but fuel. Only fuel can cause the actual weight to exceed the zero-fuel weight. This protects the wing's bending moments by placing the additional load in the wings rather than the fuselage.

To illustrate this concept, assume that you attached a three-pound weight to the center of a yard stick and then dropped the stick and the weight between two chair-backs that were placed 24 inches apart. The yard stick broke. You try it again using three, one-pound weights, one at the center and one at each end of another yard stick. It landed on the chair-backs, flexed, and remained intact. Distributing the load across the stick (wings) increased structural integrity.

EMERGENCY PROCEDURES

Memorize these procedures: engine failure, alternator failure, engine and cabin fire, inflight engine restart, manual landing gear extension, blocked static source, and the emergency descent if published. Be familiar with the remaining procedures and know where to find them.

NORMAL PROCEDURES

Never trust published fuel tank capacities.

For preflight, know where all fuel drains, fuel vents, static ports, and landing gear strut switches are located. Memorize the minimum, maximum, and recommended oil quantities. Some engines require maximum oil quantity only on very long flights. Memorize the maximum and unusable fuel quantities. If the fuel tanks have internal measuring tabs, memorize the quantities they represent. Never trust published fuel tank capacities until you have personally refueled that tank when it was empty.

Memorize the normal and hot start procedures. Are flaps recommended for takeoff? Is the electric fuel pump required for takeoffs and landings?

PERFORMANCE CHARTS

Minor temperature and altitude variations can drastically reduce airplane performance.

Pay particular attention to the gross weight takeoff and climb performance at high density altitudes. Minor temperature and altitude variations can drastically reduce airplane performance. Check the fine print on the takeoff and climb performance charts to verify whether or not flaps should be used.

Check the cruise chart for fuel consumption and memorize the highest published value. Use this value and subtract 30 minutes from the airplane's published range if you have not personally determined the airplane's actual fuel capacity and fuel consumption. Know how to use all of the POH's performance charts.

Determine the following takeoff distances:

- Maximum gross weight, sea level, temperature 15° C, and a 10 knot headwind.

• Maximum gross weight, 4,000 feet MSL, temperature 37° C, and no wind.

WEIGHT AND BALANCE

Does fuel consumption cause the airplane's CG to shift?

• Determine the airplane's center of gravity (CG) with full and with empty fuel tanks. Does fuel consumption cause the CG to shift?

• You are the pilot, a 200-pound passenger is in the copilot's seat, the remaining seats are occupied with 170-pound passengers, and the fuel tanks are full. How much baggage can be carried and where must it be located to keep the CG within limits?

• If the last problem is invalid because the pilot, passengers, and fuel exceed the useful load, compute the allowable fuel load if no baggage is carried.

• How much baggage or cargo can you load into the forward most baggage compartment if the fuel tanks are full and you are the only occupant? Check the CG carefully.

SYSTEMS

Trim

Set all trim controls to neutral or zero before preflight. During the exterior inspection, those trim tabs should be flush with the primary control surface.

If electric trim is installed, check the disconnect function before every takeoff.

If electric elevator trim is installed, know the circuit breaker location and check the disconnect function before every takeoff. Both halves of the split, electric trim switch must be moved together to operate the system. One half of the switch engages a clutch that connects the trim motor to the elevator trim cable; the other half supplies power to the trim motor.

Landing Gear

Know this system by heart. Before preflight, check that the gear handle is in the down position, turn on the master switch, and check the green, gear-down lights and the landing gear warning horn if that is possible—never raise the gear handle to check the warning horn. The gear-down lights on

Know the landing gear system by heart.

some airplanes are wired through the navigation light switch. If this switch is on during the day, the gear lights are automatically dimmed and they will appear to be inoperative.

One micro switch (called a squat switch) is normally located on one of the landing gear struts. It prevents electrical power from reaching the landing gear motor when the airplane is on the ground. Know the squat switch location, and check its condition during every preflight.

In addition, each gear strut assembly will have additional microswitches that are part of the gear warning system. They are used for the gear-down light, the unsafe or gear-up light, and the warning horn circuits. If you reduce engine power significantly or extend landing flaps when the landing gear is not extended, the warning horn will activate.

Know the location of the two landing gear circuit breakers. The low amperage breaker is for the gear warning system; the high amperage breaker is for the landing gear motor.

Most systems provide two means for confirming that the landing gear is extended.

Most landing gear systems provide two means for confirming gear extension. The primary system is the green light circuit; the secondary system is the warning horn circuit—two circuits that usually operate in parallel.

If the green light does not illuminate when the landing gear is extended, replace it with a similar light bulb—another gear-down or gear-up bulb—to determine if the bulb is burned out. If that does not fix the problem, use the gear warning horn to determine whether or not the gear is down.

There is a right way and a wrong way to use the gear horn for gear-down confirmation. First, you must know whether or not the gear light circuit and the gear horn circuit work in parallel. If they do, a valid gear-down check can be made using the gear horn.

To rely on the landing gear horn, it must be operating when the gear is extended.

Follow this mandatory sequence: With the gear handle up, reduce power until the gear warning horn activates and then extend the landing gear. If the horn stops working when the gear extends, the gear is down and locked, and you verified horn operation at the time of gear extension.

If you do not verify the horn *during* gear extension, a valid check is impossible. You erroneously assumed that the horn would work if the gear did not extend properly.

If you are unsure of the indicating system's redundancy, check with a mechanic.

Caution! If the landing gear system for the airplane you are flying does not work in the foregoing manner, study the system schematic with a mechanic in order to determine what redundancy exists for your gear indication system.

Fuel

Memorize the fuel system's schematic diagram. Know the location of all drains and vents.

Before making a hot start with a fuel injected engine, you must know how to purge air from and cool the fuel lines that are upstream of the fuel control unit in the engine compartment. If you flood the engine with excess fuel, engine oil will wash off the cylinder walls, and engine wear will increase. If the POH procedure is not satisfactory, consider the following article.

Fuel Injection System Comparisons

Excerpts from Steve Ells's article in the July 2000 AOPA Pilot magazine

Continental fuel-injection system

"The Continental fuel-injection system continually delivers a metered fuel flow to each cylinder through individual fuel-injection nozzles. Fuel is delivered from the fuel tanks to the airplane fuel selector valve. The fuel then flows through the fuel strainer and electrically driven boost pump to an engine-driven fuel pump. The engine-driven pump output pressure varies directly with the engine rpm. Fuel under pressure from the engine-driven pump is called unmetered fuel. The amount of unmetered fuel delivered to the fuel control unit is always greater than what the engine will need.

"The metered fuel, which the pilot has tailored to his flight needs with the throttle and mixture settings, is delivered from the fuel control unit to the manifold valve and then out to the fuel injection nozzles through the small stainless steel fuel lines. The excess fuel is referred to as fuel vapor (although it's generally mainly liquid) and is routed from the fuel control unit through a separate hose to the fuel tank.

"Often vapor lock will also occur in the fuel control unit; occasionally it will occur in the fuel pump. It's almost impossible to push vaporized fuel with a fuel pump—the pump is designed to push a liquid, not gaseous fumes. The result is an engine that's almost impossible to start using normal cold engine procedures.

Continental hot-start tips

"Short of waiting one to two hours for the temperatures under the cowl to cool enough so that vapor formation is no longer a problem, what should the owners of Continental fuel-injected engines do to quickly start their engines and be on their way?

"The first step is to attempt to get rid of the heat in the engine compartment while on the ground. Pointing the airplane into the wind and opening the oil filler door in the cowling will help the trapped heat on top of the engine to escape, If you are parked with the tail to the wind, opening the cowl flaps all the way *and* opening the oil cooler door will quicken cooling. Don't forget to close the oil filler door before attempting a start.

"If you are only going to be on the ground for a short time, say less than 30 minutes, this may not do the trick. One technique that I first saw printed in a Cessna Pilots Association *Technote* is worthy of note.

"This technique depends on circulating liquid fuel driven by the boost pump through the engine-driven fuel pump, the fuel control unit, and back to the fuel tank for two minutes before attempting a start. The excess fuel is returned to the tank through the fuel vapor return line.

"The theory here is that circulating liquid fuel through the engine-driven pump and fuel control unit, and returning it to the tank, will carry away the heat and vaporized fuel in the system—replacing it with cold liquid fuel that can be introduced into the cylinders for a successful start.

"Follow these simple steps: Mixture all the way out like you're trying to shut off the engine, throttle all the way in, turn the boost pump on High—and leave it there for two minutes. It may sound like the boost pump is making a lot of noise but it won't hurt the pump. After two minutes push the mixture all the way in, and watch the fuel pressure needle. When it peaks, shut off the boost pump. Pull the throttle aft to the idle setting, turn the boost pump on Low, and turn the key to start.

"The engine should start. It may be a little rough and you may need to momentarily switch the boost pump on High to purge any remaining vapor, but the engine should start within six blades of prop rotation. Lean for normal ground operation, and you are on your way. If it doesn't start, go through the fuel circulation process again—it's important to keep the fuel circulating for every second of the two minutes.

Precision Airmotive (Bendix) RSA fuel injection

"Lycoming engines are equipped with Precision Airmotive (nee Bendix) RSA fuel-injection systems. Since the RSA fuel-injection system does not have a provision to circulate the excess fuel and vapor back to the aircraft fuel tank, the Continental procedure outlined above won't work.The hot-start dilemma is further complicated by two facts. First, the amount of fuel metered to the engine in the RSA system depends upon the volume of airflow through the servo venturi and throat. During engine starts this airflow is insufficient to completely meter the proper amount of fuel.

"Also, proper metering depends on there being liquid fuel in the servo. During heat-soaked operations, the fuel in the servo is often in a vaporous state and can't actuate the internal metering system correctly. Couple this with the heat-soaked fuel manifold and steel cylinder fuel delivery lines, and hot starts can be a real headache.

Lycoming fuel-injection starting tips

"The procedure that yields dependable results is to flood the engine with fuel by pushing the mixture and throttle all the way forward and turning on the boost pump for five to 10 seconds. Then turn the boost pump off and pull the mixture to off. Turn the key to crank the engine with the throttle all the way in (open). As the engine rotates, slowly pull the throttle back toward idle.

"Closing the throttle gradually lessens the amount of air flowing to the cylinders—at some point the proper air/fuel mixture will be achieved and the engine will start to fire. When the engine starts to fire, continue to move the throttle toward idle, and be ready to gradually push the mixture control to full rich. During this procedure the engine will again start to stumble—this is the signal to push the mixture forward. The engine should then settle down into a normal idle.

"If the airplane has a manual primer, it is often helpful to squirt a couple of shots of prime into the cylinders just before attempting a start. This procedure will bypass the fuel-injection system and squirt fuel into the cylinders. It may be necessary to feed fuel to the cylinders with slow application of the primer until the fuel-injection system clears itself of vapor and begins supplying fuel."

— — — — —

Know the electric fuel pump operation. With fuel injected engines, some fuel pumps have up to three speeds each of which corresponds to a specific range of power output. For normal operations, pump speed is automatically selected by throttle position or fuel pressure. For starting, emergency, and irregular operations, pump speed is selected by the pilot.

Be prepared to turn the fuel pump on if the engine starts to die during the landing rollout.

If a fuel injected airplane does not require the fuel pump to be on for landing, be prepared to turn the pump on momentarily during the landing rollout if the engine starts to die due to fuel vaporization in the hot fuel lines. The pump can be left on if desired, but to prevent flooding, you must quickly lean the mixture until the engine runs smoothly.

Engine

Know how the induction system works. Normally, filtered air is supplied to the engine. If the filter is blocked, there must be an alternate air source. That source can be a spring-loaded alternate air door that is automatically operated by differential air pressure, a manually operated alternate air door, or carburetor heat. All three sources provide unfiltered air.

Pay particular attention to cylinder head temperature when operating high-performance airplanes. To help control temperature, cowl flaps are usually installed. They must be full open during ground operations and when flying at high power settings—climb power or greater—in moderate or warm climates. Cowl flaps are normally closed during cruise and descent, but may be partially opened if needed for cooling, a common occurrence when operating turbocharged airplanes at high altitudes. Always remember to close the cowl flaps after climbing and to open them after landing.

Upon completion of the engine run-up, check idle power.

Upon completion of the engine run-up, check idle power. Very few manufacturers list this important procedure on their checklists. If the engine will not idle, do not fly the airplane. If, however, you must fly it in order to reach a maintenance facility, do not use idle power during the landing approach.

Engine idle problems have caused many landing accidents, particularly with fuel-injected engines. The unfortunate pilot, unaware of the problem, gets high on an approach and sets idle power. Now, with no fuel flow, the residual fuel in the lines vaporizes due to engine heat. The glidepath is intercepted on short final, power is added, but nothing happens. There is not enough time to restart the engine, and the airplane lands short of the runway. Major damage is the usual result.

Know how to check the idle mixture setting.

You can easily check the idle-mixture adjustment on both carbureted and fuel-injected engines. Set idle power and slowly retard the mixture control. Initially, rpm will remain constant, but eventually the rpm should increase slightly. Further leaning will kill the engine. If rpm does not increase or if it increases excessively, the carburetor or fuel injection unit's idle-mixture setting needs adjustment. If idle speed is not correct, that too should be adjusted.

To keep the spark plugs clean while taxiing, lean the mixture.

To keep the spark plugs clean while taxiing, the ideal mixture setting is just forward of the point where rpm increased during the mixture check. If the engine hesitates when taxi power is applied, richen the mixture as necessary. Most engines will idle rough if the mixture is too rich or the engine is too cold.

The tachometer indicates engine and propeller rpm. The manifold pressure gauge indicates barometric pressure in the intake manifold, measured in inches of mercury. When the engine is shutdown, ambient air pressure will be indicated on the manifold pressure gauge—the gauge reading will be equal to the reported altimeter setting. When the engine is idling, the throttle valve is closed and pressure is low because outside air is restricted from entering the engine. At high power settings, the throttle valve opens, more air is admitted to the engine, and manifold pressure increases.

Constant-Speed Propeller

This propeller allows you to set and automatically maintain a specific rpm by using the cockpit's propeller control whenever the rpm is within the governing range. It directs the propeller governor to do all the work—change propeller pitch and rpm and then maintain that rpm.

The propeller governor is a high pressure oil pump and rpm sensor.

The governor is a high pressure oil pump and rpm sensor that sends or removes engine oil at higher-than-normal pressure to or from a moveable piston—located in the propeller hub—that changes propeller blade angle. When the propeller control is pushed forward to the high-rpm position, the oil supply is shut off from the propeller hub and the blades twist to the low-pitch, high-rpm position—they take a small bite of air. When the control is pulled aft to the low-rpm position, the maximum amount of oil is supplied to the propeller hub and the blades twist to the high-pitch, low-rpm position—they take a big bite of air.

When the propeller control is cycled between both positions, engine oil pressure will change because engine rpm is changing, and some instructors teach students to check for this while the propeller is exercised during the engine runup. That check is meaningless, because engine oil pressure always changes when engine rpm changes.

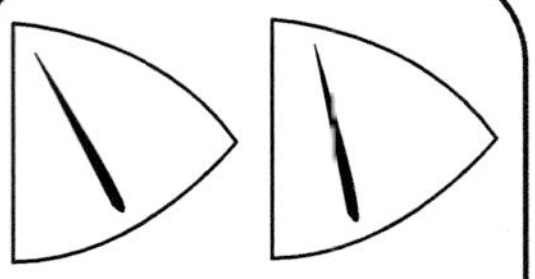

Propeller Pitch

The relationship of high rpm and low pitch versus low rpm and high pitch is often confusing. High rpm occurs when it becomes easy for the engine to turn the propeller—the blades are in the low-pitch, small-bite position. Low rpm occurs when it becomes difficult for the engine to turn the propeller—the blades are in the high-pitch, big-bite position.

During climb, cruise, and descent, the propeller control is used to set a specific rpm. If rpm starts to increase, oil pressure is increased and the blades twist toward the high-pitch, low-rpm position. If rpm starts to decrease, oil pressure is decreased and the blades twist toward the low-pitch, high-rpm position. If the engine loses oil pressure, the propeller blades will always twist to the low-pitch, high-rpm position, and the propeller control becomes inoperative. When flying in turbulence, you can hear the rpm vary as the governor tries to maintain rpm.

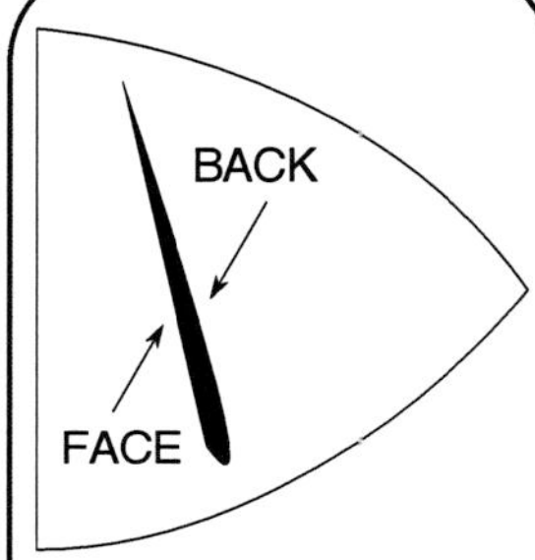

Propeller Nomenclature

The force that causes the blades to twist toward the low-pitch high-rpm position is called the centrifugal twisting moment (CTM), and it is present whenever the propeller is turning. Imagine a propeller rotating with no forward movement. The impact of the air on the face of the blade—the flat side that faces the cockpit—generates CTM that twists the blade to the low-pitch position.

The cambered side of the propeller that faces away from the cockpit is called the back of the blade. Unknowledgeable pilots call this surface the front of the blade. During preflight, pay particular attention to the leading edge and the face of each blade, the two areas where most damage occurs. In addition, always check the propeller spinner for cracks near the mounting screws.

Engine Controls

Engine damage will occur if the throttle and propeller controls are used improperly. To understand how this relationship works, think in terms of cylinder pressure.

If the throttle is increased, cylinder pressure increases because the engine is working harder. If the throttle is decreased, cylinder pressure decreases because the engine is working less.

If the propeller control is pushed forward to the high-rpm, low-pitch position, cylinder pressure decreases because the engine does not work as hard to turn the blades. If the propeller control is pulled aft to the low-rpm, high-pitch position, cylinder pressure increases because the engine must work harder to turn the blades.

In summary: High cylinder pressure—throttle forward, propeller control back. Low cylinder pressure—throttle back, propeller control forward.

Avoid high cylinder pressures.

You must avoid extreme, high pressure operation by never allowing both controls to simultaneously approach the high pressure positions. To do this, you must follow these rules when making power changes:

Adding power Move the propeller control forward and then move the throttle forward. Increasing rpm decreases cylinder pressure; increasing power increases that pressure. If the throttle is moved first, cylinder pressure may increase to an unacceptable level.

Reducing power Move the throttle back and then move the propeller control back. Reducing power decreases cylinder pressure; reducing rpm increases that pressure. If the propeller control is moved first, cylinder pressure may increase to an unacceptable level.

Leaning the mixture is a mandatory procedure.

The mixture control is used to improve engine efficiency, increase fuel economy and range, eliminate engine roughness, extend spark plug life, and eliminate fouling. The rules for operating the mixture control during power changes are:

Adding power Richen the mixture before you add power.

Reducing power Lean the mixture after you reduce power.

Unless the POH states otherwise, engines should be leaned when the power output drops to 75 percent or less. Always do this during climb, cruise, and descent. When landing, richen the mixture to the takeoff power setting that the airport's density altitude would require—full rich at sea level.

To manually lean an engine in flight, retard the mixture until roughness occurs. Now, advance the mixture until the engine smooths out, note that increment of forward movement, and advance the mixture until that distance has doubled. For example, you retard the mixture until the engine starts to run rough. When you then advance the mixture for smooth operation, the control moved forward 1/4 inch, so move it forward another 1/4 inch. The engine is now leaned close to the best power setting.

Over leaning may cause engine damage.

If the airplane is equipped with an exhaust gas temperature (EGT) gauge, it is easy to keep the mixture leaned. For best power, most procedures instruct you to lean to peak EGT and then richen the mixture 50 degrees; for best economy, lean to peak EGT—use caution when the EGT probe is connected to only one cylinder. Over leaning may cause engine damage, because with only one EGT probe, there is no guarantee that the hottest cylinder is being monitored.

EGT gauges have a movable pointer or an adjustable reference mark that can be set at the 50-degree rich-of-peak position. When power changes are made, adjust the mixture control to keep the EGT needle at that reference mark.

Turbocharger

The turbocharger is a device with two vaned wheels—a compressor and a turbine—bolted to each end of a common shaft. It uses engine exhaust gas to spin the turbine which causes the compressor to spin and produce high pressure ambient air for the engine's air-intake system. Compressor output varies directly with the amount of exhaust gas that flows through the turbine, and that gas is controlled by either a fixed or a variable waste gate—a diverter valve that is located in the exhaust manifold upstream from the turbine. Fixed waste gates have no controls, but most variable waste gates have automatic controls that maintain the desired manifold pressure.

The term critical altitude defines the point where the turbocharger can no longer supply sea-level air pressure and power output begins to decrease. At this point, the waste gate is fully closed and all of the engine's exhaust gas is used to spin the turbine wheel.

With turbochargers it is quite easy to exceed manifold pressure and temperature limits.

The primary precaution for operating turbocharged engines is to keep cylinder pressures and temperatures within limits. Observe manifold pressure limits and do not allow an engine to be overboosted. Observe cylinder head temperatures and do not allow the engine to become overheated.

For best economy a turbocharged airplane must be flown at high altitudes, and if the airplane is unpressurized, all occupants must breath supplemental oxygen. Do not conduct high altitude flights until you have received high altitude physiological training in a high-altitude decompression chamber. Call your local FAA Flight Standards District Office for training locations.

Flying turbocharged airplanes above the critical altitude requires attentiveness and careful power management. At high altitude the engine responds to anything that changes the mass flow (fuel-air charge) that enters the induction system. Airspeed, manifold pressure, EGT, fuel flow, and rpm all become interrelated, and if one of these factors or the outside air temperature changes, the other factors will change.

Avionics

If you are not certain how an advanced avionics unit works, turn it off and leave it off.

High performance airplanes usually have advanced avionics equipment. Many items are not required for visual flying, and if you positively know how to turn off equipment that you're not proficient with, you can fly the airplane. If, however, you want to use that equipment, study the operating manual and obtain an equipment checkout from a knowledgeable flight instructor.

Check the autopilot disconnect function before every takeoff.

Autopilots demand particular attention. Know the location of the autopilot circuit breaker and be certain to test the disconnect function before every takeoff. Find out whether the attitude indicator or the turn coordinator provides reference information for the autopilot, a question that is never answered in the airplane's POH.

The autopilot's reference source is critical knowledge for instrument flying, because you must know which instrument failure will cause autopilot failure. The same can be said for visual night flying, a close parallel to instrument flying when flying over areas with minimal lighting.

AIRWORK

The transition to complex airplanes will be enjoyable if you were properly trained as a private pilot. Otherwise, your instructor will be forced to spend considerable time developing your awareness and teaching basic procedures, maneuvers, and attitude reference points. The thought of flying these airplanes is exciting, but without the proper skills, complex airplanes can be deadly unforgiving.

During your checkout, you will be asked to fly using instrument reference. If you are unable to maintain positive airplane control while executing 180-degree turns, climbs, and descents, additional instrument instruction will be required.

In the long run, an instrument rating is a wise investment.

Complex airplanes and cross-country flying go hand-in-hand. When flying cross-country, it is easy to get into situations that require instrument reference—hazy sky conditions and dark nights over unlighted terrain. Such environments frequently require instrument reference; in the long run, an instrument rating is a wise investment.

PITCH PERFORMANCE

Chapter 3 discussed and illustrated pitch performance. That illustration is reproduced and expanded on the following page in order to show you the approximate power settings for basic, complex training airplanes.

Both illustrations depict the same five P, I, T, C, H attitudes. Apply this knowledge during your checkout, and you will quickly realize why I took the time to write this manual.

All wings work in the same way.

Since you have been taught to use the wing as a primary reference, you'll feel right at home flying any airplane once you learn the new power settings. All wings work in the same way; you will prove this to yourself each time you check out in a different type of airplane.

Remember this important point: If you set the proper attitude and power setting, where will the airspeed be? Within five or six knots of the desired value. You are safe, and the airplane is achieving the basic performance objective.

During your checkout you will make slight pitch attitude adjustments in order to obtain the exact airspeed. Now you are flying more precisely, but at no time were you flying in an unsafe manner. This is an important distinction that leads to the best part of this entire concept—your self-confidence is always high; your performance is always outstanding.

DEPARTURES

Retract the landing gear when a positive climb occurs and fly the departure profile as described in chapter 4 and reviewed on the next page. You identify a positive climb by looking out

Attitude	Performance Objective	Power Setting
P 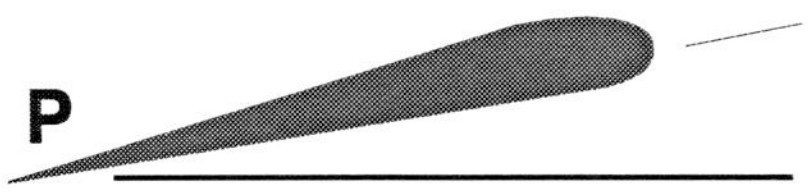	Best Angle Climb	Max. Continuous
I 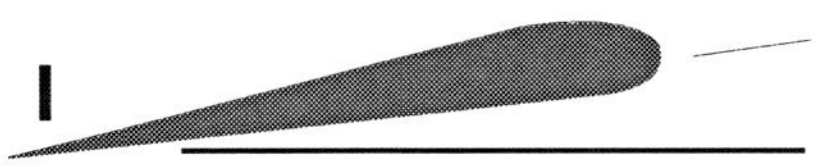	Best Rate Climb	Max. Continuous
T 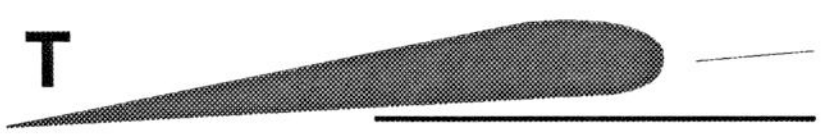	Cruise Climb Slow Cruise Gear Down Gear Up	Climb 2,000-2,200 rpm, or 20-22 Inches MP 17-19 Inches MP
C 	Accelerate Cruise Best Glide Slow Cruise Descent Gear Down Gear Up Power-On Landing Approach	Takeoff Cruise Idle 1,400-1,500 rpm, or 14-15 Inches MP 12 Inches MP 1,300-1,400 rpm, or 13-14 Inches MP
H 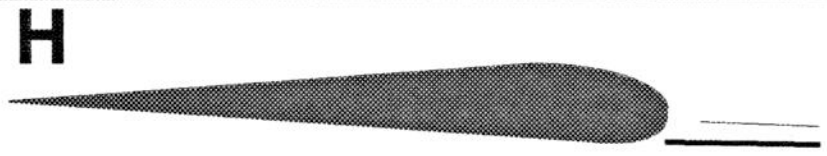	Cruise Descent Power-Off Landing Approach	Cruise Idle

Pitch Performance

Retract the landing gear when a positive climb occurs.

the window at the pie-in-the-sky sight picture and evaluating airplane movement in relation to the ground. Instrument pilots who depart in low visibility conditions must use the vertical speed indicator to confirm a positive rate of climb, and that is where the expression "positive rate, gear up" comes from—contrary to popular belief, it does not apply to visual flying.

Climb to a safe altitude as rapidly as practical.

To leave the landing gear extended so that a landing can be made on the remaining runway if the engine fails is academic nonsense. A smart pilot always accelerates to Vy as fast as possible and climbs at that speed to a safe altitude. Then, and only then, is power reduced.

Try this exercise at an airport with a long runway: Take off and leave the gear down. Climb 100 feet, chop the power, and land. You will be amazed at the amount of runway that is required for this exercise.

Your personal safety is the name of the game.

Your safety is the primary objective. You must always be able to walk away from the airplane calm, cool, and collected, regardless of what went wrong. If the engine fails just after takeoff, fate has played you a dirty trick. After all, your pre-departure checks were satisfactory, all checklists were completed, and all engine indications were normal at the start of the takeoff roll. So, you have two options:

- Attempt to get back on the ground with the gear down and stop the airplane on the runway. If you choose to do this, you are playing test pilot because of all the variables—wind; landing impact; tire, brake, and runway condition; and your alertness. Light airplane brakes are marginal at best, and if you depart the runway, structural damage can easily occur and jam the cabin door shut. If a fuel leak were to occur, a serious fire hazard exists and you may be trapped in your coffin.

- With the landing gear retracted, you make a belly landing on the runway, slide 200 feet, open the door, and walk off the airplane with your passengers safe and sound.

I will always take option number two and so should you. The airplane slapped me in the face, and I'm slapping her back. "All's fair in love and war," and this situation is war.

DEPARTURE PROFILE

Execution

Clear the final approach for traffic, and taxi onto the runway. Maintain a level pitch attitude during the initial acceleration, and establish the cruise-climb attitude as soon as the elevator becomes effective. Maintain this attitude while the airplane lifts off the runway and accelerates to the best-rate climb speed (V_Y). When a positive climb occurs, retract the landing gear.

Just before you anticipate hearing the *clunk* sound of the wheels entering the wheel wells, establish the best-rate climb attitude and maintain it until you reach a safe altitude. Now re-establish the cruise-climb attitude, accelerate to that speed, and set climb power.

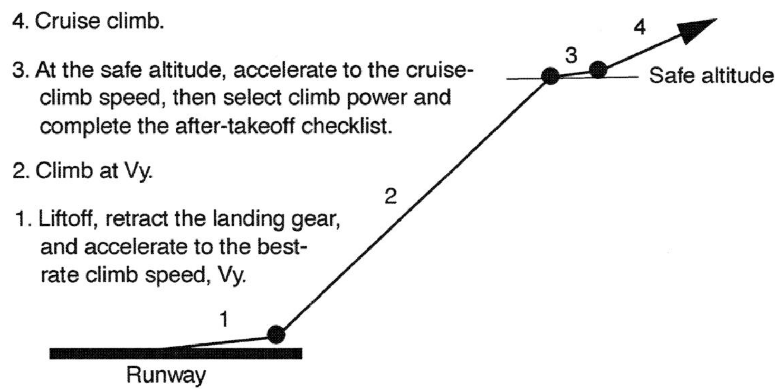

The Departure Profile

Attitude

At the start of the takeoff roll, maintain the C-pitch attitude for initial acceleration. As elevator pressure becomes effective, establish and maintain the T-pitch attitude while the airplane accelerates, lifts off, and accelerates to V_Y.

To maintain the T-pitch attitude and accelerate rapidly to V_Y, you must move the yoke forward slightly as airspeed increases. If elevator position is held constant, the airplane will pitch up at liftoff, as airspeed increases, and when climbing out of ground effect. Landing gear retraction may also cause an undesired pitch change.

Approaching V_Y, establish and maintain the I-pitch attitude until you reach the safe altitude, and then re-establish the T-pitch attitude for the cruise climb. Learn to sense the approach of V_Y by feeling elevator pressure and listening to propeller noise. With constant speed propellers, engine rpm (noise) remains constant, but control pressures and propeller blade-tip noise increase as airspeed increases.

SHORT-FIELD TAKEOFF

Execution

Clear the surrounding airspace for traffic. As described by the departure profile and the following adjustments:

- Set flaps as recommended by the airplane's POH, taxi into position at the absolute end of the runway, set brakes, set takeoff power, release brakes, and begin the takeoff roll. Use the C- and T-pitch attitudes in the normal manner.

- At liftoff, instead of maintaining the T-pitch attitude, positively increase pitch to the P-pitch attitude for best angle climb. Stop increasing pitch if an intermittent stall warning or prestall buffet occurs.

- When clear of the 50-foot obstacle, establish the T-pitch attitude, retract the gear, accelerate to V_Y, retract the flaps, and follow the departure profile.

Drag usually increases during landing gear retraction.

Landing gear retraction is delayed until you clear the obstruction and select the T-pitch attitude for acceleration, because landing gear door operation and the geometry of many landing gear systems during the retraction process will increase drag and decrease performance. If this is not the case for the airplane you are flying, you can retract the gear when a positive climb occurs.

SOFT-FIELD TAKEOFF

Execution

Clear the surrounding airspace for traffic. As described by the departure profile and the following adjustments:

- Set takeoff flaps, taxi onto the runway with the yoke held full back, do not stop, and smoothly select takeoff power.

- Let the nose rise to the P-pitch attitude as soon as possible, and maintain that attitude until liftoff occurs. Stall warning will occur while accelerating to the liftoff point. If necessary, decrease pitch attitude slightly to keep the main wheels on the ground while the stall warning is activated.

Stay close to the runway until you establish the T-pitch attitude.

- At liftoff, decrease pitch attitude to avoid an excessive climb, and stay close to the runway while the airplane accelerates. When pitch decreases to the T-pitch attitude, maintain that attitude, allow the airplane to climb, and retract the gear when a positive climb occurs. When the airplane accelerates to V_Y, select the I-pitch attitude, retract the flaps, and follow the departure profile.

DESCENTS AND APPROACHES

Sudden cooling is detrimental to the life of air-cooled engines.

Due to engine thermal shock—sudden cooling that is detrimental to the life of air-cooled engines—idle or low power settings in high performance airplanes are not recommended. Idle power is acceptable during the landing flare since the airplane is at slow speed. Check the POH for a minimum recommended power setting, and if nothing is published, use 13 to 15 inches of manifold pressure. This minimum power setting should keep the engine driving the propeller rather than allowing a windmilling propeller to drive the engine, a situation that you can immediately hear.

Idle or low power settings, as listed on page 9 - 15's pitch performance chart, are acceptable in basic, complex training airplanes, but for advanced high-performance airplanes, these power settings are unacceptable. To obtain the required descent performance and keep the power up, increase drag as necessary.

Your first action to increase drag is the extension of approach flaps—normally, the first increment of flaps. If more drag is required, extend the landing gear. If airspeed must remain above the approach-flaps extension speed, your only option is to keep the flaps up and extend the landing gear.

LANDINGS

Every year, pilots make gear-up landings in complex airplanes. To avoid this fate, always follow this rule:

GEAR DOWN, GO DOWN

In other words, use the landing gear rather than the throttle to start the descent from traffic pattern altitude. This procedure is compatible with the principles of CAAPVAA—the constant attitude and power, variable airspeed approach.

Always triple-check your gear down indications.

The second item in the GUMPS mental checklist—Undercarriage—reminds you to check the gear-down indications. You must do this three times: when descending below pattern altitude, turning onto final approach, and cleared to land.

ARRIVAL PROFILE

Execution

As described on page 5 - 21 and reviewed here in order to add landing gear and propeller operation. Clear the surrounding airspace for traffic. Start your descent from cruise altitude, but you must do it earlier than normal when flying faster airplanes. If groundspeed is approximately three miles per minute (180 knots) and you descend at 500 fpm, you will travel six nautical miles for every 1,000 feet of descent.

The propeller control is always placed in the high-rpm, low-pitch position prior to landing.

To descend from pattern altitude, extend the landing gear and then reduce power if a higher descent rate is necessary. To prepare for a possible go-around, smoothly move the propeller control to the high-rpm position. Pilots who say, "I'll do that on short final," usually forget it when told to go-around.

As stated in chapter 5, you make only four power reductions between cruise altitude and landing: a reduction for the cruise descent if required to keep airspeed within limits or for turbulence, a reduction to slow cruise when leveling off at traffic pattern altitude, a reduction to increase the descent rate if necessary after you extend the landing gear, and a reduction to idle power just prior to touchdown. Flying in this manner makes you look very smooth and very sharp. Practice this at every opportunity.

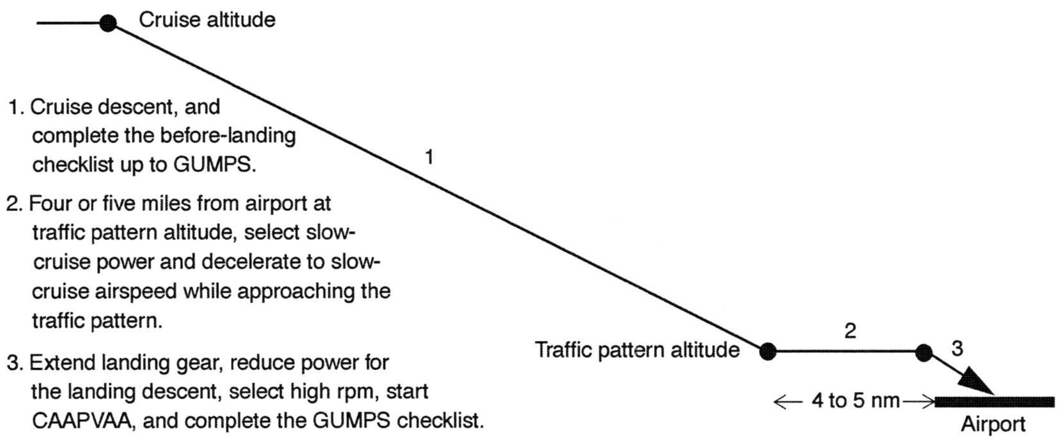

The Arrival Profile

EMERGENCY DESCENT

Maximum landing gear extended speed (V_{LE}), landing gear extended, flaps up, and a 45-degree bank unless the airplane's POH specifies otherwise.

EMERGENCY APPROACH AND LANDING

Additional glide distance can be obtained by moving the propeller control to the low-rpm, high-pitch position.

If the engine fails at altitude in a complex airplane, proper propeller and landing gear operation must be considered. Best glide still uses the C-pitch attitude and the high-rpm, low-pitch propeller position, but you can obtain additional glide distance if you move the propeller control to the low-rpm, high-pitch position. When you do this, you can feel the airplane accelerate as propeller drag decreases. Save this option for insurance just in case you inadvertently get too low on the approach.

Ideally, the gear is extended on the downwind leg abeam the planned touchdown spot, but you can extend the gear sooner or later depending on your altitude. You may, however, be required to extend the landing gear manually, an action that can take two or three minutes in some airplanes.

GO-AROUNDS

Murphy's Law: "Anything that can go wrong will go wrong."

During go-arounds, particularly at night, think twice before you simultaneously retract electrically powered landing gear and electrically powered wing flaps. With both systems operating, a high load is placed on the alternator. The electrical system is designed for these loads, but remember Murphy's Law, which always occurs at the worst time.

To be prepared for a possible go-around, you must set the propeller control to the high-rpm, low-pitch position after the landing gear has been extended and the power has been reduced for the descent from traffic pattern altitude. To go-around, set takeoff power, select approach flaps, and retract the landing gear when a positive climb occurs.

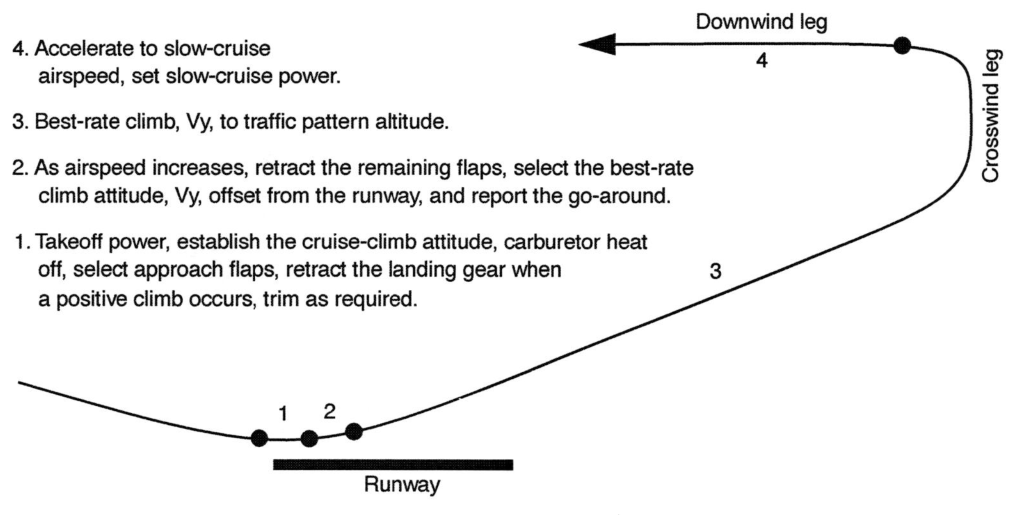

The Go-Around

INSIGHTS (More of my articles from the AOPA's *Flight Training* magazine)

Haste Makes Waste

Avoid foolish mistakes - December 2003 issue

My first encounter with a complicated airplane was quite an education; my enthusiastic responses to simulated problems nearly did me in. When a fault light illuminated or a gauge's needle jumped beyond acceptable limits, I reacted impulsively. That action was a big mistake.

The situation improved when I learned to sit on my hands for ten seconds after observing an abnormality. Quick decisions and flailing hands were replaced by intelligent evaluations and appropriate actions. This lesson should be taught during general aviation flight training.

Aircraft abnormalities are divided into two groups: emergencies and irregularities. An emergency requires both immediate and reference action; an irregularity requires timely reference action. Immediate action items are performed from memory. Reference action items are performed using a written checklist.

Engine failure and aircraft fire are emergencies. A rough engine, fuel flow fluctuations, landing gear problems, electrical malfunctions, instrument failure, or an open cabin door are irregularities.

Evaluation is the first step, if the problem is not obvious. What happened, and can you prove your assumption? For example, engine oil pressure has dropped to zero. Before you declare an emergency or execute a forced landing, check oil temperature. If it's normal, the oil pressure gauge is probably the problem. Temperature will climb if pressure is dropping.

Instrument training provides a good opportunity to teach the importance of proper evaluation. During partial panel flight—the attitude and heading indicators are covered—your instructor says, "I smell electrical smoke."

When you start to turn off avionics and other electrical equipment to isolate the problem, your instructor says, "Why don't you just turn off the master switch in order to immediately stop the fire's origin and then turn off the appliances?"

If you follow your instructor's advice without thinking, he will ask, "How long can you continue to fly on instruments?" Don't say, "Until I run out of gas," because the electric turn coordinator—an instrument that you must have in this situation—lost power when you turned off the master switch.

Validation is the second step. If you move a switch, you must confirm that the desired action occurred. For example, if you turn a fuel pump on or off you must observe a change in fuel pressure.

A landing gear indicator problem provides an example of proper validation. Most pilot operating handbooks (POHs) are vague on this subject, but prudent decisions require a complete understanding of the landing gear system.

Retractable gear airplanes usually have two gear-warning systems that work in parallel: the horn and the lights. If a gear-down light does not illuminate after the gear is extended and swapping light bulbs does not help, the warning horn can be used to confirm landing-gear extension. But, a hazard may still exist if the horn failed since it was last tested. To validate the horn, raise the landing gear and climb to a safe altitude, retard the throttle until the horn activates, and extend the gear while the horn is blaring. If the horn shuts off, you have a valid gear-down indication.

Many factors can generate similar traps: the lack of cockpit familiarity, system knowledge, or sleep; high workload; turbulence; and poor cockpit lighting. Can you identify every switch and control in your cockpit while blindfolded? Are you mentally prepared to cope with all conceivable problems in that airplane?

Instead of flying the airplane every time you get in it, spend some time sitting in it on the ground with the POH and think about possible failures. If the information is limited, ask a mechanic for help. With proper knowledge you can avoid situations that prove the old adage, "haste makes waste."

Positive Climb, Gear Up

Use common sense - August 2003 issue

People resist change, particularly when a previous belief—right or wrong—is held with total confidence. When I learned to fly, my confidence was directly related to that displayed by my instructor. When another instructor, pilot, or publication disagreed with what I had been taught, my initial reaction was denial. How could I be wrong? That attitude was a big mistake.

In retrospect, some training procedures conflicted with practicality. What I learned made sense, but it was refuted by experienced pilots. Also, the complexity of some ground school material often ignored the importance of simple procedures that are mandatory in a properly managed airplane.

Flight instructors easily fall into this trap. If most of their flight experience has been acquired in a training environment, their intent, logic, and self-confidence may be genuine, but they may be espousing incorrect techniques. Landing-gear retraction after takeoff is a perfect example. One of three techniques is normally taught for landing-gear retraction: when a landing on the runway is no longer possible; when the vertical speed indicator shows a positive rate of climb; or when a positive climb is evident while looking outside the cockpit.

To evaluate the first technique, add your airplane's no-wind takeoff and landing distances to and from 50 feet agl. A Cessna 182RG at maximum gross weight requires about 2,800 feet; a Cessna 210, 3,600 feet. These figures are conservative because they are based on maximum short-field performance and do not consider your engine-failure reaction time. Above 50 ft., the landing distance increases dramatically. At 100 feet AGL, I estimate an additional 2,500 feet for each airplane.

If you land with the gear down after engine failure, other factors will increase landing distance: skidding, improper tire inflation, tire slippage on the wheel rim, brake ineffectiveness due to wing lift, and high brake temperatures—a common occurrence when a pilot drags the brakes to control excessive taxi speed. If none of these factors applies, you should still realize that light-airplane brakes are far from ideal when a high-speed, maximum-effort stop is required.

The greatest hazard is a runway excursion—departing the end or side of the runway—because of a blown tire, wheelbarrowing, or an incorrect estimate of landing distance. Subsequent structural damage could jam the cabin door shut, and if a fuel tank leak and fire occur, your airplane may become your coffin. Aircraft structural integrity becomes a myth when an aircraft collides with objects on the ground or in the air.

The second technique requires a positive-rate confirmation before gear retraction. This is an instrument-flying requirement for low-visibility takeoffs; it is not a visual flying requirement. In this case, the vertical speed indicator is used for climb confirmation.

The third technique is correct for visual flying. At liftoff, keep your eyes outside the cockpit in order to maintain the existing pitch attitude so that the airplane will rapidly accelerate to V_Y, the best rate of climb speed. When you recognize a positive climb—it's obvious that you're climbing—retract the gear. When it is almost retracted—just before the usual clunk sound—establish the airplane's V_Y pitch attitude. Now, glance at the airspeed indicator. Amazing! It will be very close to V_Y—the quickest way to reach the altitude that gives you a suitable forced-landing option.

If the engine does quit immediately after or during gear retraction—an extremely rare occurrence—consider this: You determined that the airplane was airworthy during the preflight inspection, the pretakeoff checks were satisfactory, and all engine parameters were normal after takeoff power was applied. You have just been hijacked by your airplane.

Leave the gear up, put the airplane back on the runway, and slide to a stop. Only a few hundred feet are required for a belly landing. Now, open the cabin door and walk off unscathed with your passengers.

In other words, the airplane slapped you in the face, so you must slap her back. Your safety and that of your passengers is your first priority, not the airplane.

Dismay

Stupid is as stupid does - January 2004 issue

You may not be flying a complex airplane at this time, but sooner or later most pilots do make this transition. All pilots, regardless of their experience level, should be acutely aware of the pilot errors that repeatedly lead to gear-up landings.

These accidents often involve pilots who cannot accept time-proven procedures. They modify such procedures or make additions in order to make themselves standout among their peers or students. Unfortunately such actions usually increase the complexity of flying, ignore the realities of human performance, and decrease pilot self-confidence.

Checklists are the first consideration. Knowledgeable pilots use flow patterns and mental checklists such as the following to complete specific tasks: CIGARS before takeoff—controls, instruments, gas, attitude trim, runup and radios, and seatbelts—and GUMPS before landing—gas, undercarriage, mixture, prop, and seatbelts. Then, with the exception of GUMPS, they check their performance by using a written checklist.

Pilots and instructors who add other items to these mental checklists are making a big mistake, because these short, easy to remember mnemonics are designed to be the basic elements that prevent accidents and aircraft damage. They were never intended to include a myriad of other actions that are meaningful "house-keeping items" but have no impact on flight safety. CIGARS and GUMPS comply with the KISS principle—keep if simple stupid!

GUMPS is not backed-up with a written checklist. It follows the before-landing written checklist, which should be reviewed before you enter the traffic pattern where maintaining a traffic watch is essential. Upon completion, you can then say that the before-landing checklist is complete with the exception of GUMPS, because the landing gear has not yet been extended, the mixture may not yet be set for the airport's altitude, and the propeller has not been placed in the low-pitch position.

Landing gear extension is the next consideration. You must follow the axiom of *gear down, go down.* A descent from pattern altitude should never start with a power reduction in a retractable-gear airplane; it should start with the drag that occurs when the landing gear is extended. Landing-gear extension signals the first of three requirements for the GUMPS checklist. The second GUMPS requirement occurs when you are cleared to land; the third occurs when you are on short-final. This rule-of-three will keep you from making an embarrassing mistake.

It is impossible to mandate one procedure for all situations, but you must develop the habit of using GUMPS three times. Humans make mistakes, and good pilots learn to triple-check critical actions. Show me someone who thinks he or she is God's gift to aviation and beyond such thinking, and I'll show you an accident that's waiting to happen.

Exceptions for the gear-down, go-down philosophy occur when you are high due to ATC altitude restrictions, terrain, or poor descent planning; or if you are forced to follow a slower airplane. In such cases, the landing gear must be extended early. Exceptions also occur when you are cleared for a straight-in or base-leg-entry approach.

These exceptions do not excuse you from triple checking your performance, because flight safety is not an exception, it is mandatory. Perform the GUMPS checklist when you pass through pattern altitude (usually 1,000 feet above ground level) when cleared to land, and when on short final.

Avoiding gear-up landings does not require more rules and regulations. It does require common sense, awareness, and good judgment. Too often we see the results of pilots who, for one reason or another, temporarily ignore these attributes—even though they possess them and sometimes even teach them. Every pilot and every flight instructor must keep basic flight safety procedures simple and make them habitual.

REVIEW TEST

1. What commitment must you make in order to stay current in complex airplanes?
2. What speeds can you adjust as weight decreases and what is the rule for those adjustments?
3. If you encounter moderate turbulence, what is the quickest way to decelerate to maneuvering speed?
4. What happens to the allowable G-Load when you extend flaps in a normal category airplane?
5. After takeoff, should you always reduce power for the climb?
6. Can six-place airplanes always carry six passenger and full fuel?
7. When you checkout in a complex airplane, what CG consideration should you check?
8. Before preflighting the airplane, what should you do with the trim controls?
9. What is your major concern with a retractable landing gear system?
10. During preflight, should you raise the landing gear selector in order to check the landing gear warning horn?
11. If you use the gear horn to confirm gear extension due to a failure of the normal indicating system, what is the proper way to it?
12. To hot start a fuel injected engine you must?
13. During engine run-up, particularly with fuel injected engines, what is the one item that you should always check even though it is not listed on most checklists?
14. If a fuel injected engine starts to die during the landing rollout, you should?
15. What high-performance airplane component is used to control engine temperatures?
16. When the engine is shutdown, what should the manifold pressure gauge read?
17. What does the manifold pressure gauge indicate?
18. How does manifold pressure change with throttle changes?
19. What happens when you move the propeller control between the high rpm and the low rpm positions?
20. Define the face and the back of a propeller blade.
21. How does the constant-speed function of the propeller work?
22. If the engine loses oil pressure, what will a constant-speed propeller do?
23. What is the correct control action for adding power?
24. What is the correct control action for reducing power?
25. What can happen if you use the engine controls improperly?
26. Describe a turbocharger.
27. What does critical altitude mean?
28. Where should you go before you fly above 10,000 feet in turbocharged airplanes?

29. What must you know about the autopilot and the electric trim before you takeoff?
30. Why are low power settings not recommended for high performance airplane engines?
31. Define a positive climb.
32. After takeoff, when do you retract the landing gear?
33. What is your most important consideration if an engine fails on takeoff?
34. Why do you delay landing gear retraction until the obstacle is cleared on a short-field takeoff?
35. What is engine thermal shock?
36. With respect to landing gear extension, what procedure prevents gear-up landings?
37. Before landing, when must you check the gear-down indications?
38. Where do you position the propeller control prior to landing and why?
39. When you execute a go-around, what precaution should you observe?
40. What common reference can you rely on for flying simple, fixed-gear airplanes and complex, high-performance airplanes?

"SUPERIOR PILOTS STAY OUT OF TROUBLE BY USING THEIR SUPERIOR JUDGMENT TO AVOID SITUATIONS THAT WOULD REQUIRE THE USE OF THEIR SUPERIOR SKILLS."

author unknown

APPENDIX

The following Skyroamers® checklists for visual flight were written for Cessna 152s and 172s and Piper Warriors and Archers that were produced prior to 1998. The checklists follow chapter 2's standard operating procedures and reinforce the concept of using flow patterns, mental checklists, and written checklists. The basic checklist format can be adapted to any airplane that you use for visual flight training.

"THERE ARE OLD PILOTS AND
THERE ARE BOLD PILOTS, BUT
THERE ARE NO OLD, BOLD PILOTS."

author unknown

CESSNA 152, 172 VFR CHECKLIST

PREFLIGHT

Announce "Complete" when finished

Initial Exterior Inspection

Fuel/Oil/Tires/Windshield - CHECK
Servicing - CALL NOW if required
Tiedowns/Covers - REMOVE
Doors/Compartments - UNLOCK
Ignition Key - SET ON GLARESHIELD

Power-off Cockpit Check

First, use flow pattern to check that everything is OFF or CLOSED.

Documents - CHECK (AROW)
Hobbs Meter - CHECK TIME
Elevator Trim - NEUTRAL
Electrical Switches/Radios - OFF
Controls - UNLOCKED

Power-on Cockpit Check

Battery Switch - ON
Fuel Gauges - CHECK
Flaps - EXTENDED
Lights - CHECK if required
Battery Switch - OFF

Exterior Inspection

Per Pilot's Operating Handbook

Tires - CHECK while moving airplane

BEFORE START

Announce "Complete" when finished

First, use flow pattern to prepare airplane for engine start.

Seats - LOCKED in position
Safety Belts - FASTENED
Fuel Selector - ON/BOTH
Elevator Trim - SET for takeoff
Mixture - SET for engine start
Throttle - 1/2 inch OPEN
Carburetor Heat - OFF
Circuit Breakers - IN
Parking Brake - OFF
Battery and Alternator Switches - ON
Alternator Low Voltage Light - ON
Primer - USE if needed and LOCKED
Radio - ON for ATIS, then OFF
Altimeter - SET
Brakes - SET
Area - CLEAR and callout "CLEAR"
Starter - ENGAGE

AFTER START

RPM - 1,000
Oil Pressure - CHECK
Flaps - UP
Mixture - LEANED for taxi
Rotating Beacon - ON
Radio Master Switch - ON
HI - SET
Brakes - CHECKED

BEFORE TAKEOFF (CIGARS)

Announce "Complete" when finished

While Taxiing

Controls - FREE, CORRECT MOVEMENT
Instruments:
  MC - TURNING, HEADING CHECKED

In Runup Area

Gas:
  Fuel Selector - ON/BOTH
Attitude trim:
  Elevator - SET
Runup:
  Area - CLEAR
  Parking Brake - ON
  Mixture - SET for takeoff
  Throttle - 1,700 rpm

CESSNA VFR CHECKLIST

Suction Gauge - CHECKED
Ammeter - CHECKED with load
Carburetor Heat - CHECKED
Magnetos - CHECKED (125/50)
Oil press. and temp. - CHECKED
Throttle - IDLE CHECK, then1,000 rpm

Radios:
Comm/Nav - SET FOR DEPARTURE
Transponder - SET

Safety Belts, Doors, Windows - SECURED

TAXIING ONTO RUNWAY

LIGHTS - ON (Taxi, Landing, Strobes)
CAMERA - Transponder on ALT
ACTION - Time, Oil Pressure, RPM

AFTER TAKEOFF

Announce "Complete" when finished

Landing and Taxi Lights - OFF
Oil Press. and Temp. - CHECKED

BEFORE LANDING

Announce "Complete" when finished

Safety Belts - FASTENED
HI - SET
Altimeter - SET
Fuel Selector - ON/BOTH
Mixture - SET for go-around
Oil Pres. and Temp. - CHECKED
Landing and Taxi Lights - ON

AFTER LANDING

LIGHTS - OFF as required
CAMERA - Transponder OFF
ACTION - Trim, Flaps, Mixture leaned for taxi, Carb Heat

ENGINE SHUTDOWN

Announce "Complete" when finished

ELT - CHECK on 121.5
Radios and Electrical Equipment - OFF
Throttle - IDLE
Magnetos - GROUND CHECK
Throttle - 1,000 to 1,200 rpm
Mixture - IDLE CUTOFF
Throttle - CLOSED
Battery and Alternator Switches - OFF
Magnetos - OFF, remove key and set on glareshield.
Fuel Selector - OFF/LEFT or RIGHT
Tach/Hobbs Meter - RECORD TIMES
Maintenance Squawks - RECORD

PARKING

Battery Switch - RECHECK OFF
Parking Brake - OFF
Control Lock - INSTALLED
Trash - REMOVED
Doors/Compartments - LOCKED
Propeller - VERTICAL if required for fuel
Wheel Chocks - IN PLACE
Tie Downs - SECURED
Flight Plan - CLOSED

PIPER WARRIOR, ARCHER VFR CHECKLIST

PREFLIGHT

Announce "Complete" when finished

Initial Exterior Inspection

Fuel/Oil/Tires/Windshield - CHECK
Servicing - CALL NOW if required
Tiedowns/Covers - REMOVE
Doors/Compartments - UNLOCK
Ignition Key - SET ON GLARESHIELD

Power-Off Cockpit Check

First, use flow pattern to check that everything is OFF or CLOSED.
Documents - CHECK (AROW)
Hobbs Meter - CHECK TIME
Electrical Switches/Radios - OFF
Controls - UNLOCKED
Elevator Trim - NEUTRAL
Flaps - EXTENDED

Power-on Cockpit Check

Master Switch - ON
Fuel Gauges - CHECK
Lights - CHECK if required
Master Switch - OFF

Exterior Inspection

Per Pilot's Operating Handbook
Tires - CHECK while moving airplane

BEFORE START

Announce "Complete" when finished

First, use flow pattern to prepare airplane for engine start.
Seats - LOCKED in position
Safety Belts - FASTENED
Elevator Trim - SET for takeoff
Flaps - UP
Rudder Trim - SET for takeoff
Parking Brake - ON
Fuel Selector - LOWEST TANK
Battery Switch - ON
Radio - ON for ATIS, then OFF
Alternator Switch - ON
Altimeter - SET
VAC, ALT, OIL Warning Lights - ON
Circuit Breakers - IN
Carburetor Heat - OFF
Fuel Pump - ON, check pressure
Mixture - SET for engine start
Throttle - 1/2 inch OPEN
Primer - USE if needed and LOCKED
Area - CLEAR and callout "CLEAR"
Starter - ENGAGE

AFTER START

RPM - 1,000
Oil Pressure - CHECK
Mixture - LEANED for taxi
Rotating Beacon - ON. OFF if strobes are on same switch.
Fuel Pump - OFF, check pressure
HI - SET
Brakes - CHECKED

BEFORE TAKEOFF (CIGARS)

Announce "Complete" when finished

While Taxiing

<u>C</u>ontrols - FREE, CORRECT MOVEMENT
<u>I</u>nstruments:
MC - TURNING, HEADING CHECKED

In Runup Area

<u>G</u>as:
Fuel Selector - SWITCH TANKS
<u>A</u>ttitude trim:
Elevator and Rudder - SET
<u>R</u>unup:
Area - CLEAR
Parking Brake - ON

PIPER VFR CHECKLIST

Mixture - SET for takeoff
Throttle - 2,000 rpm
Oil press. and temp. - CHECKED
Ammeter - CHECKED with load
Vacuum Gauge - CHECKED
Carburetor Heat - CHECKED
Magnetos - CHECKED (125/50)
Throttle - IDLE CHECK, then1,000 rpm whenever stopped
Fuel Pump - ON

Radios:
Comm/Nav - SET FOR DEPARTURE
Transponder - SET

Saftey Belts, Doors, Windows - SECURED

TAXIING ONTO RUNWAY

LIGHTS - ON (Taxi, Landing, Strobes)
CAMERA - Transponder on ALT
ACTION - Time, Oil Pressure, RPM

AFTER TAKEOFF

Announce "Complete" when finished

Landing and Taxi Lights - OFF
Fuel Pump - OFF
Oil Press. and Temp. - CHECKED

BEFORE LANDING

Announce "Complete" when finished

Safety Belts - FASTENED
HI - SET
Altimeter - SET
Fuel Pump - ON
Fuel Selector - SET
Mixture - SET for go-around
Oil Pres. and Temp. - CHECKED
Landing and Taxi Lights - ON

AFTER LANDING

LIGHTS - As required
CAMERA - Transponder OFF
ACTION - Trim, Flaps, Mixture leaned for taxi, Carb Heat, Fuel Pump

ENGINE SHUTDOWN

Announce "Complete" when finished

ELT - CHECK on 121.5
Radios and Electrical Equipment - OFF
Throttle - IDLE
Magnetos - GROUND CHECK
Throttle - 1,000 to 1,200 rpm
Mixture - IDLE CUTOFF
Throttle - CLOSED
Master Switch - OFF
Magnetos - OFF, remove key and set on glareshield.
Tach/Hobbs Meter - RECORD TIMES
Maintenance Squawks - RECORD

PARKING

Master Switch - RECHECK OFF
Parking Brake - OFF
Control Wheel - SECURED with seat belt
Trash - REMOVED
Doors/Compartments - LOCKED
Propeller - VERTICAL if required for fuel
Wheel Chocks - IN PLACE
Tie Downs - SECURED
Flight Plan - CLOSED

PRIVATE PILOT TESTING

A practical test for a private pilot certificate is a simple, fun-filled, no-hassle endeavor. Correct? No way!

Testing is a pain in the neck for most of us, but one fact is undeniable: You feel great after some exams, but incredibly lucky if you pass others. Assuming that you were prepared, what caused those opposite reactions? The examiner.

Did he put you at ease at the start, or were you jerked to the edge of your seat on the first round of questions? Did the questions concentrate on the information that you were expected to know, or did the examiner try to determine how much you knew? Those factors will determine how you felt after the test.

I'll never forget my early testing experiences. In those days all tests were given by FAA inspectors—old guys with lots of general aviation experience. If a three-hour oral was scheduled, it took all of three hours, but it was usually a rewarding experience.

The easy questions came first, and subsequent questions were straight forward and dealt with the required material. When it was obvious that I was properly prepared, most examiners would say, "The exam's over, you've passed, but I'd like to ask you some other questions."

Knowing that he had my undivided attention, he then turned our session into a learning experience. As the loaded questions began—questions with implications that could easily generate an incorrect response—I stumbled, he hinted, and I learned.

This approach to testing reveals the secret for test preparation. Be rock solid on the basic material and stick to that material if the examiner drifts into gray areas before you know whether or not you've passed the test. Without practical experience, you can't be the master of a subject, and because you're at the testing level, you obviously lack practical experience. Stick to basic facts that you know are correct.

Don't be dead serious during the test. You can really help yourself if you use humor whenever it's appropriate. That trick usually reduces the initial tension that always exists when you are tested.

For example, when my students get in the airplane with an examiner at the start of a flight test, I tell them to say, "I must brief you on the operation of the safety belts, but with your experience you know how they work. Is there anything else that you'd like to know?"

Hopefully the examiner will say, "Well, what else would you like to tell me?"

The student should then say, "I want your signature and I want you out of this airplane." That will always bring forth a laugh. The student will then say, "Seriously, help me look out for traffic, and when I ask you test your brakes after we start to taxi, don't jam them on, because that will upset our passengers." Ha-ha!

You have no control over an examiner who makes judgment evaluations based on hypothetical situations. You can, however, bring everything to a halt when the questioning becomes vague or unrealistic. Make the examiner be specific.

Some people who spend all of their time checking students become bored and look for new twists to old material. Unfortunately, that confuses most students. If you ever become an examiner, never lose sight of what it was like to be a student. Don't complicate simple concepts, and never drift away from required material in order to present hypothetical situations that require a high level of practical experience.

I once knew an instrument instructor who did just that. He had a system for setting up two VOR receivers so that he could fly direct to an intersection from an off-airway location. It was interesting, but unfortunately he was teaching a procedure that was absolutely never used in actual practice until the advent of GPS navigation.

If this instructor was an examiner, what do you think might happen? Obsessed with his discovery, he might present it as a question, knowing full well that it would be impossible for the student to know the answer. During the attempt to derive an answer, the student might make an erroneous statement about VOR operation that would be disqualifying. In the end, the solution would be revealed, but would that clarification be made to satisfy the student or the examiner's ego?

When asked how to prepare for the private pilot practical test, I refer students to the FAA's *Private Pilot Practical Test Standards*. Page 3 lists twenty-one study references. When you know that material, you will easily pass the oral test.

During the flight test, you should verbalize your thoughts so that the examiner knows what you're thinking. The best way to pass a flight test is to keep the examiner in the information loop.

Self-confidence is your most important consideration. It must be high. Never go into an test if you have doubts about your ability to succeed. After you complete the practical test, you will receive your private pilot certificate.

Congratulations! Now you can advance to the next learning plateau—putting what you have learned into practice.

"PILOTS MUST BE WARY, AND THEY MUST
HAVE AN INFINITE CAPACITY FOR SUSPICION.
ON THE GROUND, AWAY FROM AIRPORTS,
YOU MAY BE AN OPTIMIST, BUT AT
AN AIRPORT AND IN AIRPLANES
YOU MUST BE A PESSIMIST."

author unknown

TEST ANSWERS

Chapter 2

(1) Formalized rules that deal with safety, awareness, and professionalism. (2) The pilot's operating handbook. (3) Opposite to the normal direction of rotation. (4) Fire, unless the starter was cranking the engine. (5) Lean the mixture. (6) Retard the mixture until the engine runs rough. Then advance the mixture twice the distance that's required for smooth operation. (7) The rpm will decrease for fixed-pitch propeller engine. (8) Full heat unless a carburetor air temperature gauge is available. Full throttle is also required for maximum heat, but if considerable ice has already formed, moving the throttle may dislodge the ice and cause the engine to stop running. (9) The idle check after the engine runup. It must be satisfactory, particularly with fuel injected engines. (10) Magnetos grounded and emergency locator (ELT) not transmitting. (11) Flow patterns and mental checklists backed up with the written checklists. (12) The written checklist. (13) State who you are calling, who you are, where you are to include altitude when airborne, and what you want. (14) No! (15) Occupants removed and grounding wire connected between the airplane and the fuel truck. (16) Vertical, not circular. (17) Deformed tanks and aircraft attitude. (18) Wind conditions, airport traffic flow, congestion around the airplane, and the best way to exit the tie-down area. (19) You the pilot. Are you fit to fly? (20) When on the runway and when airborne. Do not use them while taxiing or waiting for takeoff. (21) When operating in high density airspace or within 10 miles of an airport, and whenever you deem it necessary to make your airplane more conspicuous. (22) Use the lowest tank for engine start and initial taxi. Use the fullest tank (opposite tank if both are full) for engine run-up and takeoff. (23) Select idle power with the throttle before you turn the magneto switch back on. (24) The brakes—make a smooth brake application, but do not stop the airplane. Use the good brake to stop the airplane while maintaining directional control with the rudder pedals. (25) The pilot taking control states, "I have it." The pilot relinquishing control states, "You have it." (26) Select idle power and use spot braking. Never drag the brakes to control taxi speed. (27) The magnetic compass heading when taxiing parallel to or perpendicular to a runway; and the directional gyro, attitude indicator, and turn coordinator. (28) Never! Use the tow bar and avoid blasting other airplanes or open hangars. (29) Chain to rigid pavement anchor: interlocking links method. Chain to flexible anchor cable: hook method if permitted. Rope: locking half-hitch method. (30) Four seconds inside, sixteen seconds outside. (31) Absolutely none. (32) Always have an out. (33) By keeping track of time aloft and known fuel consumption. Pilots never rely on the fuel gauges. (34) Fixating—concentrating on one object, one problem, or one task.

Chapter 3

(1) Vision for attitude and relative motion, hearing for airspeed and engine speed, touch for control pressures, kinesthesia for motion, and smell for odors. (2) The airplane's orientation in relation to the Earth's surface. Rotational: pitching, rolling, and yawing. Linear: horizontal, lateral, and vertical. (3) Refer to the illustrations on page 3 - 4 and 3 - 6. (4) Refer to the table on page 3 - 7. (5) Ailerons: returned to the faired position. Rudder: returns to a position that maintains the proper yaw rate, seldom the faired position. Elevator: returns to a position that maintains the proper pitch attitude, seldom the faired position. (6) First, make a correction that is obviously excessive. When it starts to work, remove half of that input. (7) First, make a correction that will stop the deviation from increasing. Second, make a correction that will eliminate the deviation. (8) You keep track of power changes by using measured increments of hand movement. (9) Stabilize engine power and airspeed, use the elevator to maintain pitch attitude, and use elevator trim to remove elevator control pressure. (10) Stabilize engine power and airspeed, trim the elevator, hold the wings level, use rudder to maintain heading, and use rudder trim to remove rudder control pressure. (11) You get "behind" the airplane, lose your control input awareness, and increase your workload. (12) After you trim the airplane, you let go of the yoke and control the airplane with only rudder and throttle. Maintain the desired bank angle with rudder; stop slight climbs or descents with slight power changes. (13) Horizontally so that you can easily apply full rudder input. Seat-back tilt so that you are comfortable when holding the yoke. Vertically so that you just barely see the top of the engine cowl while looking over the glare shield. (14) A reduced cruise airspeed normally located just within the airplane's flap extension speed. It is used for the traffic pattern's entry and downwind legs and when cruising in the practice area between maneuvers. (15) Clear your airspace for traffic, and announce it verbally to your instructor. (16) Evaluate rate of turn in rela-

tion to bank angle and airspeed. The inclinometer is a secondary reference. (17) Entering a climb: If flying at climb speed, simultaneously set climb power and establish the climb attitude. If flying faster then climb speed, establish the climb attitude, but delay the power increase until airspeed decreases to climb speed. For the level-off, establish the desired attitude, but do not reduce power until airspeed has increased to the desired cruise speed. (18) Entering a descent: If flying at the descent speed, simultaneously set descent power and establish the descent attitude. If flying faster then descent speed, set descent power, maintain altitude until airspeed decreases to descent speed, and then establish the descent attitude. For the level-off, simultaneously establish the desired attitude and set the desired power setting. (19) To maximize forward speed in order to save time and money. (20) The climb that generates the highest vertical speed. The quickest way to reach a higher altitude. (21) Constant airspeed: Your hands simultaneously move in opposite directions. Constant attitude: Your hands move in the same direction. (22) To learn the airplane's personality. (23) Decreasing pitch attitude and adding power. (24) Aggressive rudder input. (25) If a wing rises, step on it with rudder input. (26) No change. Only groundspeed changes. (27) Turns: varying bank angle. Straight flight: varying heading. (28) To maintain a given distance from a rectangular reference. (29) To fly symmetrical half circles that lie on opposite sides of a road. (30) To nullify wind drift during a crosswind approach and landing. (31) To rapidly lose altitude without increasing airspeed. (32) So that the steepest bank of the maneuver will occur when it is started.

CHAPTER 4

(1) To reach a safe altitude as fast as possible. Refer to page 4 - 2. (2) The lowest altitude that provides a suitable landing site in the event of engine failure. (3) Maintain directional control with rudder. When the elevator becomes effective, select the T-pitch attitude, maintain it until airspeed increases to Vy, and let the airplane fly itself off the ground. (4) Increased elevator moment arm, leaving ground effect, and increasing airspeed. (5) Excess fuel is being pumped into the engine for valve cooling. This would not occur if a reduced power setting was used for the climb. (6) When 300 feet below pattern altitude. (7) The view point: check for traffic on the entry and downwind legs. The check point: apply carburetor heat, check engine oil pressure and temperature, review the GUMPS mental checklist, check for traffic ahead in the pattern, and call the tower for landing clearance. The key point: begin the landing evaluation. (8) When you must reduce power in order to start the CAAPVAA. (9) Start at the runway and work backward in the traffic pattern. (10) It forces you to maintain your forced-landing approach proficiency, and it is a self-improvement maneuver—the more you do it, the better you get at evaluating the involved factors. (11) The turn to base leg, base leg orientation, flap selection, the turn onto final approach, and your ability to anticipate vertical air movement and wind gradient changes. (12) Square off the turn to final so that you can use a steeper than normal bank angle. This increases the normal ground track and sink rate. (13) Execute a forward slip. (14) Go around! (15) Evaluate surface features and sun position to determine areas of sinking and rising air. As altitude decreases, wind velocity decreases and wind direction shifts counterclockwise. (16) From full flap extension to the flare point. From the flare point to touchdown. From touchdown until reaching taxi speed. (17) Stage 1: Airspeed, attitude, and the forward sight picture for the touchdown spot and wind drift. Stage 2: Attitude and the pie-in-the-sky sight picture, for height, sink, and drift. Stage 3: The pie-in-the-sky sight picture until the nose wheel touches down. You can then use the forward sight picture in order to maintain directional control as you decelerate to taxi speed. (18) Wheelbarrowing and loss of control. (19) Ailerons to stay over the runway centerline; rudder to stay parallel with the centerline. (20) After you slow to taxi speed. (21) Refer to page 4 - 21. (22) A go-around is a normal maneuver that reflects good judgment. Radio communications. (23) Outside the cockpit, never inside when traveling at high speed on the ground. (24) Add power, maintain the C-pitch attitude until elevator pressure feels normal, and then select the T-pitch attitude and regain altitude. (25) Make a large power increase and select the T-pitch attitude. When sink rate is normal, select the C-pitch attitude, but do not decrease power until elevator pressure feels normal. (26) Select the C-pitch attitude and add power to prevent further airspeed decrease. Descend to the proper height, resume flaring, and when the I-pitch attitude is established, slowly reduce power and let the airplane settle onto the runway. (27) Add power. Select the I-pitch attitude, slowly reduce power, and let the airplane settle onto the runway. (28) Close the throttle, maintain the landing attitude with elevator and directional control with rudder. (29) Maintain present pitch attitude and add enough power to prevent further airspeed decay. As you approach the runway, resume flaring, establish the I-pitch attitude, slowly reduce power, and let the airplane settle onto the runway. (30) Decrease pitch attitude to establish a normal sink rate and add considerable power to prevent further airspeed decay. If a safe landing is in doubt, go around. (31) Quickly move the yoke aft to increase pitch attitude. As the nose pops up, remove crab with rudder, stop drift, force the upwind wheel onto the runway with aileron, and maintain pitch attitude with the elevator. (32) Relaxing elevator back pressure which can lead to wheelbarrowing and loss of control; forgetting to retract the flaps, reposition the elevator trim, or turn off the carburetor heat; and if an error in judgment occurs, aborting the takeoff within the confines of the runway may be impossible. (33) You glance inside the cockpit when traveling at high speed on the runway. The flaps and carburetor heat controls are moved while on the runway. The flaps are retracted before full power is selected. (34) To use good judgment and comply with your instructor's instructions.

CHAPTER 5

(1) The magnetic compass. (2) Acceleration and deceleration errors, and the northerly turning error. (3) The most altitude with the least forward travel. The steepest climb. (4) A power increase as the airplane rolls through 30 degrees of bank, a G-load increase when the airplane reaches 45 degrees of bank, and removal of both when you start the recovery. (5) By varying the bank angle ± 5 degrees. (6) Yes, you can in most modern training airplanes, but you must use the rudder to maintain bank attitude not the ailerons.

(7) Yes. (8) Skidding: An accelerated stall followed by a spin entry toward the inside of the turn. Slipping: An accelerated stall followed by a spin entry toward the outside of the turn. (9) Throttle closed, ailerons neutralized, full opposite rudder, neutralize the rudder when the spin stops, yoke forward to break the stall and recovery from the ensuing dive. (10) No, but it could enter a high speed, descending spiral. (11) Steepest bank occurs when flying directly downwind at the highest groundspeed. Shallowest bank occurs when flying directly upwind at the slowest groundspeed. (12) When flying with a direct crosswind. Minimum crab angle (zero) occurs when flying upwind or downwind. (13) So that the first turn will have the steepest bank. If that bank angle is too steep—more than 45 degrees—restart the maneuver further from the pylon. (14) Yes, because it is a prolonged turn. (15) Surface wind indicators. (16) Ailerons are displaced at the beginning of the takeoff roll. The T-pitch attitude selection is delayed. Rudder is used to stay parallel with the runway centerline, and ailerons are used to stay over the centerline until a positive climb becomes apparent. (17) Rudder is used to stay parallel with the runway centerline. Ailerons are used to stay over the centerline. When the upwind main wheel touches down, increase the aileron input. Pitch attitudes are identical with those used for normal landings. (18) The temptation to lower the nose and force the airplane onto the runway which will cause wheelbarrowing. Go around and start over. (19) You confirm that you are positively climbing away from the runway by looking outside the cockpit, not by looking at the altimeter or vertical speed indicator. (20) The C- and T-pitch attitudes in the normal manner. At liftoff, the P-pitch attitude. When clear of the obstacle, the T-pitch attitude for acceleration to Vy which is maintained by using the I-pitch attitude in conjunction with the departure profile. (21) Select the P-pitch attitude as soon as possible. At liftoff, select the T-pitch attitude at a rate that keeps the airplane close to the runway. As the airplane approaches Vy, capture the I-pitch attitude in conjunction with the departure profile. (22) To smoothly transition from cruise altitude to the traffic pattern in minimum time, with minimum power changes, and to maximize the opportunity for traffic observation while approaching the airport. (23) About 35 miles [(7.5 X 2 X 2) + 5 = 35]. (24) Bank the airplane to remain over the runway centerline. When the upwind main wheel touches down, apply additional aileron input in order to hold the upwind wheel on the runway. Apply full aileron input in strong wind conditions. (25) Keep the airplane parallel to the runway centerline. (26) A steeper than normal, straight-line glidepath that extends down final approach to the touchdown spot and just clears the obstacle. (27) After engine failure, best glide will maximize your glide distance, minimum sink will maximize your time aloft. (28) ABC. A—fly the Airplane. B—pick the Best landing site. C—Cockpit check for a possible engine restart. (29) The altitude loss for that airplane during a 225-degree, power-off turn. (30) A precautionary landing.

CHAPTER 6

(1) Attitude, proficiency, and knowledge. (2) Pilotage, landfall, and dead-reckoning navigation. (3) In the center of the chart's airport symbol for the departure and destination airports. Don't try to second guess ATC. (4) At least every 10 miles. (5) Your groundspeed in order to validate the winds aloft forecast that you based your flight time and fuel consumption on. (6) Magnetic course. You need to know it in order to determine the proper cruise altitude. (7) East is least and west is best. (8) Refer to the chart on page 6 - 7. (9) Watch the Weather Channel's weekly planner. (10) You are validating the forecast. If actual weather is equal to or better than forecast, you can place a high level of confidence in that forecast and plan your flight. (11) Yes. If you have already received an abbreviated briefing, you want to check for any last minute TFRs and weather changes along your route of flight? (12) Increase the published fuel consumption by approximately ten percent, and then compute the total flight time based on that fuel consumption and the published fuel capacity. Now subtract 30 minutes from the total flight time, and then subtract your reserve fuel requirement from that value in order to determine the time that is available for your planned flight. (13) Traffic, attitude, wind evaluation, and locating your reference points. (14) Time. (15) The finger method. (16) Monitoring your groundspeed, updating your en route and destination weather and making pilot reports if appropriate with en route flight advisory service, monitoring HIWAS VOR stations for hazardous weather advisories, and utilizing ATC's radar advisory service if possible for traffic advisories. (17) Whether or not that airport has a special departure procedure. (18) A taxiway and runway diagram for that airport. (19) Return to the last reference point and try again. (20) The four-Cs: climb, communicate, confess, and comply. (21) No. (22) 318 (23) About 21 minutes. You will fly 11 miles every 6 minutes and 5.5 miles every 3 minutes (6 + 6 + 6 + 3 = 21). (24) Approximately 100 knots. 42 minutes plus 18 (about 1/2 of 42) equals 60 minutes. 67 miles plus 33 (about 1/2 of 67) equals 100 miles. (25) 116 knots. 11% (2 x 5.5) of 105 is approximately 11. 105 + 11 = 116. (26) 36 knot headwind component (0.9 X 40). 20 knot crosswind component (0.5 X 40). (27) About 3.5 minutes.

CHAPTER 7

(1) Neither will move. (2) Like heading, like bearing. (3) Tune the frequency, identify the station, and set the OBS. (4) No, the transmitted signal or the receiver has failed. (5) It will not move. It remains pointed toward the station. The compass card is turning beneath the needle. (6) Parked. (7) When orienting your position to a station. (8) Select the ADF position and tune a station. Deflect the needle from the tuned position using the test switch or by tuning a second station. When the test switch is released or the original station is retuned, the needle should snap back to the original position. (9) The ADF position utilizes the sense antenna and the loop antenna which allows the course needle to operate. The REC or ANT position utilizes only the sense antenna, and the course needle is inoperative. (10) Tune the frequency, identify the station, and set the radio compass card to the parked position. (11) The radio compass card does not have an OFF flag. You must always listen to the station identifier to make certain that a valid signal is being received. (12) Station orientation. Tune and identify the station. Center the CDI with a From indication. The OBS now indicates the radial you are on, the bearing from the station. (13) Station orientation. Tune and identify the station and set the airplane's magnetic heading at

the top of the radio compass card. The tail of the ADF needle now indicates the bearing from the station. (14) Yes. (15) The CDI is centered. (16) Select the cross-bearing on the OBS. If the CDI points toward that VOR, the cross-bearing is ahead of you. If the CDI points away from that VOR, the cross-bearing is behind you. (17) Homing results in a curved ground track to the station; tracking compensates for wind, which results in a straight ground track. (18) Turn 20 degrees toward the course. (19) Remove half of the intercept angle. (20) Turn back to the original intercept heading. (21) Turn back to the no-wind heading, and let the wind drift you back on course. (22) You can "chase the needle," but you would not turn more than five degrees. (23) The To-From flag changes to From. (24) The course needle which has been pointing toward the airplane's nose rotates and points toward the airplane's tail. (25) The VOR receiver's OBS OFF flag appears; the NDB station's identifier stops transmitting.

CHAPTER 8

(1) It performs an internal self test and an initialization test. (2) Four. (3) A function that ensures an accurate position. (4) Five. (5) Activating the wrong input, or activating the correct input at the wrong time. (6) Leg and OBS. (7) OBS. (8) The active waypoint that you enter into the GPS receiver—an airport, a VOR or NDB, an intersection, a latitude/longitude coordinate, or a bearing and distance from any of those references. (9) Spending too much time with your eyes inside the cockpit. (10) Because only three hours are required, whereas it takes 12 to 15 hours to master basic instrument flying for an instrument rating. (11) Establish straight-and-level flight at slow-cruise power, adjust the AI, set the HI, apply carburetor heat, and turn on the pitot heat. (12) Refer to the illustration on page 8 - 8. (13) Watch the AI and roll into a 15-degree bank. When established scan the altimeter and the HI. (14) Add full power and establish a wing's-level, two-bar nose high attitude. Adjust pitch for airspeed. When stabilized look at the AI and roll into a 15-degree bank. When established scan the AI, airspeed indicator, and HI. (15) Reduce power three- or four-hundred rpm, and establish a wing's-level, one-bar nose low attitude. Adjust pitch for airspeed. When stabilized look at the AI and roll into a 15-degree bank. When established scan the AI, airspeed indicator, and HI. (16) Simultaneously add power, decrease pitch attitude to the level flight attitude, and level the wings. (17) Simultaneously reduce power and level the wings. Now increase pitch to the level flight attitude. (18) Light control pressures and slow attitude changes. (19) FIrst, put in a correction that you know is excessive. Second, when it becomes apparent that the correction is starting to work, remove half of the initial input. (20) FIrst, stop the incorrect trend that caused the deviation. Second, establish a new trend that will eliminate the deviation. (21) Don't fly with a weak airplane battery. (22) Forced landing options. (23) Always carry two with spare batteries, and never place a flashlight on the glareshield close to the magnetic compass. (24) When on a taxiway at night and other aircraft are in close proximity, unless you turn them on momentarily to warn other aircraft of your presence. (25) Closer. (26) High. (27) Closer. (28) Farther.

CHAPTER 9

(1) The U.S. Army says four hours a month. (2) Best glide, stall, maneuvering, and final approach speeds. For every two percent below gross weight, reduce those speeds approximately one percent. (3) Reduce power, increase pitch, and if necessary, extend the landing gear at V_{LO}. (4) Decreases well below 3.8 Gs. (5) No. Refer to the POH. (6) No. (7) How the CG changes as fuel is consumed. (8) Set them to zero. During preflight, the tabs should then be faired with the primary control surface. (9) To fully understand the indicating system's redundancy. (10) Absolutely not. (11) The gear horn must be operating when you move the landing gear selector to the down position, and it must shut off when the landing gear extends. (12) Purge the air from and cool the fuel lines upstream of the fuel control unit. (13) Idle. (14) Momentarily activate the electric fuel pump. (15) Cowl flaps. (16) Current ambient air pressure. (17) Actual barometric pressure in the intake manifold. (18) Opening the throttle lets more air into the engine and manifold pressure increases; closing the throttle restricts air from entering the engine and manifold pressure decreases. (19) When moved to the high-rpm position, CTM moves the blades to the low-pitch position; when moved to the low-rpm position, engine oil pressure moves the blades to the high-pitch position. (20) Face: The rear surface that faces the airplane. Back: The forward surface. (21) A speed governor senses rpm changes. If speed increases, oil pressure increases pitch; if speed decreases, oil pressure decreases and CTM decreases pitch. (22) CTM will move the blades to the low-pitch, high-rpm position. (23) Mixture rich, rpm increased, and throttle increased. (24) Throttle decreased, rpm decreased, and mixture leaned. (25) Excessive cylinder pressures. (26) Two vaned wheels mounted on a common shaft. The turbine wheel is driven by exhaust gases, and It causes the compressor wheel to turn, which forces outside air into the engine's induction system. (27) The altitude where a turbocharged engine can no longer generate sea-level power. (28) A high altitude chamber. (29) That the autopilot disconnect button works and the location of the autopilot and electric-trim circuit breakers. (30) The adverse effects of thermal shock. (31) It is obvious from looking outside the cockpit that the airplane is climbing. (32) As soon as a positive climb becomes apparent. (33) Your personal safety. (34) With most complex airplanes, drag increases and performance decreases while the gear is retracting. (35) Sudden engine cooling, a situation that reduces engine life. (36) Gear down, go down. (37) Three times: leaving pattern altitude, turning onto final approach, and when cleared to land. (38) High-rpm position so that you can immediately set takeoff power if a go-around becomes necessary. (39) Do not operate electrically powered landing gear and flaps simultaneously. (40) The wings' P-I-T-C-H and B-A-N-K attitudes.

INDEX

Continued on next page

Continued on next page

Continued on next page

Continued on next page

Continued on next page